232
Advances in Polymer Science

Advances in Polymer Science

Recently Published and Forthcoming Volumes

Biopolymers
Vol. 232, 2010

Polymer Materials
Vol. 231, 2010

Polymer Characterization
Vol. 230, 2010

Modern Techniques for Nano- and Microreactors/-reactions
Volume Editor: Caruso, F.
Vol. 229, 2010

Complex Macromolecular Systems II
Volume Editors: Müller, A.H.E., Schmidt, H.-W.
Vol. 228, 2010

Complex Macromolecular Systems I
Volume Editors: Müller, A.H.E., Schmidt, H.-W.
Vol. 227, 2010

Shape-Memory Polymers
Volume Editor: Lendlein, A.
Vol. 226, 2010

Polymer Libraries
Volume Editors: Meier, M.A.R., Webster, D.C.
Vol. 225, 2010

Polymer Membranes/Biomembranes
Volume Editors: Meier, W.P., Knoll, W.
Vol. 224, 2010

Organic Electronics
Volume Editors: Meller, G., Grasser, T.
Vol. 223, 2010

Inclusion Polymers
Volume Editor: Wenz, G.
Vol. 222, 2009

Advanced Computer Simulation Approaches for Soft Matter Sciences III
Volume Editors: Holm, C., Kremer, K.
Vol. 221, 2009

Self-Assembled Nanomaterials II
Nanotubes
Volume Editor: Shimizu, T.
Vol. 220, 2008

Self-Assembled Nanomaterials I
Nanofibers
Volume Editor: Shimizu, T.
Vol. 219, 2008

Interfacial Processes and Molecular Aggregation of Surfactants
Volume Editor: Narayanan, R.
Vol. 218, 2008

New Frontiers in Polymer Synthesis
Volume Editor: Kobayashi, S.
Vol. 217, 2008

Polymers for Fuel Cells II
Volume Editor: Scherer, G.G.
Vol. 216, 2008

Polymers for Fuel Cells I
Volume Editor: Scherer, G.G.
Vol. 215, 2008

Photoresponsive Polymers II
Volume Editors: Marder, S.R., Lee, K.-S.
Vol. 214, 2008

Photoresponsive Polymers I
Volume Editors: Marder, S.R., Lee, K.-S.
Vol. 213, 2008

Polyfluorenes
Volume Editors: Scherf, U., Neher, D.
Vol. 212, 2008

Chromatography for Sustainable Polymeric Materials
Renewable, Degradable and Recyclable
Volume Editors: Albertsson, A.-C., Hakkarainen, M.
Vol. 211, 2008

Wax Crystal Control · Nanocomposites Stimuli-Responsive Polymers
Vol. 210, 2008

Functional Materials and Biomaterials
Vol. 209, 2007

Phase-Separated Interpenetrating Polymer Networks
Authors: Lipatov, Y.S., Alekseeva, T.
Vol. 208, 2007

Biopolymers

Lignin, Proteins, Bioactive Nanocomposites

Volume Editors: Akihiro Abe
Karel Dušek
Shiro Kobayashi

With contributions by

S. Blazewicz · H. Hatakeyama · T. Hatakeyama
A. Ikai · K. Pielichowska

Springer

Editors
Akihiro Abe
Professor Emeritus
Tokyo Institute of Technology
6-27-12 Hiyoshi-Honcho, Kohoku-ku
Yokohama 223-0062, Japan
aabe34@xc4.so-net.ne.jp

Shiro Kobayashi
R & D Center for Bio-based Materials
Kyoto Institute of Technology
Matsugasaki, Sakyo-ku
Kyoto 606-8585, Japan
kobayash@kit.ac.jp

Karel Dušek
Institute of Macromolecular Chemistry
Czech Academy of Sciences
of the Czech Republic
Heyrovský Sq. 2
16206 Prague 6, Czech Republic
dusek@imc.cas.cz

ISSN 0065-3195 e-ISSN 1436-5030
ISBN 978-3-642-26461-0 ISBN 978-3-642-13630-6 (eBook)
DOI 10.1007/978-3-642-13630-6
Springer Heidelberg Dordrecht London New York

Cover design: WMXDesign GmbH, Heidelberg

Printed on acid-free paper

Springer is part of Springer Science+Business Media (www.springer.com)

Volume Editors

Akihiro Abe

Professor Emeritus
Tokyo Institute of Technology
6-27-12 Hiyoshi-Honcho, Kohoku-ku
Yokohama 223-0062, Japan
aabe34@xc4.so-net.ne.jp

Karel Dušek

Institute of Macromolecular Chemistry
Czech Academy of Sciences
of the Czech Republic
Heyrovský Sq. 2
16206 Prague 6, Czech Republic
dusek@imc.cas.cz

Shiro Kobayashi

R & D Center for Bio-based Materials
Kyoto Institute of Technology
Matsugasaki, Sakyo-ku
Kyoto 606-8585, Japan
kobayash@kit.ac.jp

Editorial Board

Advances in Polymer Sciences Also Available Electronically

Advances in Polymer Sciences is included in Springer's eBook package *Chemistry and Materials Science*. If a library does not opt for the whole package, the book series may be bought on a subscription basis. Also, all back volumes are available electronically.

For all customers who have a standing order to the print version of *Advances in Polymer Sciences*, we offer the electronic version via SpringerLink free of charge.

If you do not have access, you can still view the table of contents of each volume and the abstract of each article by going to the SpringerLink homepage, clicking on "Browse by Online Libraries", then "Chemical Sciences", and finally choose *Advances in Polymer Science*.

You will find information about the

- Editorial Board
- Aims and Scope
- Instructions for Authors
- Sample Contribution

at springer.com using the search function by typing in *Advances in Polymer Sciences*.

Color figures are published in full color in the electronic version on SpringerLink.

Aims and Scope

The series *Advances in Polymer Science* presents critical reviews of the present and future trends in polymer and biopolymer science including chemistry, physical chemistry, physics and material science. It is addressed to all scientists at universities and in industry who wish to keep abreast of advances in the topics covered.

Review articles for the topical volumes are invited by the volume editors. As a rule, single contributions are also specially commissioned. The editors and publishers will, however, always be pleased to receive suggestions and supplementary information. Papers are accepted for *Advances in Polymer Science* in English.

In references *Advances in Polymer Sciences* is abbreviated as *Adv Polym Sci* and is cited as a journal.

Special volumes are edited by well known guest editors who invite reputed authors for the review articles in their volumes.

Impact Factor in 2009: 4.600; Section "Polymer Science": Rank 4 of 73

Publisher's Note for Volumes with Independent Contributions

This volume was not originally planned as a topical volume by the stated Volume Editors. It contains substantial review articles on interesting and valuable subjects which were invited and reviewed by the stated Editors on an independent basis and then – upon receipt – put together in one volume by the Publisher.

Marion Hertel
Executive Editor Chemistry, Springer

Contents

Adv Polym Sci (2010) 232: 1–63
DOI:10.1007/12_2009_12

Published online: 26 August 2009

Lignin Structure, Properties, and Applications

Hyoe Hatakeyama and Tatsuko Hatakeyama

Abstract Polymeric features of lignin and its potential as a bio-resource are reviewed, focusing on its characteristic structure and properties. Lignin is a random copolymer consisting of phenylpropane units having characteristic side chains. Lignin slightly crosslinks and takes an amorphous structure in the solid state. The molecular motion is observed as glass transition by thermal, viscoelastic and spectroscopic measurements. The hydroxyl group of lignin plays a crucial role in interaction with water. By chemical and thermal decomposition, a wide range of chemicals can be obtained from lignin that can be used as starting materials for synthetic polymers, such as polyesters, polyethers, and polystyrene derivatives. At the same time, a variety of polymers can be derived from lignin by simple chemical modification. The hydroxyl group acts as a reaction site for the above chemical reaction.

Keywords Biocomposites · Glass transition · Lignin · PCL · Polyurethanes

Contents

H. Hatakeyama
Graduate School of Engineering, Fukui University of Technology, 3-6-1, Gakuen, Fukui 910-8505, Japan

T. Hatakeyama (✉)
Lignocel Research, 73-8, Yatsumata, Fukui 910-3558, Japan
e-mail:lignocel@mx3.fctv.ne.jp

Abbreviations

ρ	Apparent density
σ	Breaking strength
σ_{10}	Strength at 10% strain
τ_c	Correlation time
ADL	Acetylated DL
AFM	Atomic force microscopy
BDF	Biodiesel fuel
b-NMR	Broad line nuclear magnetic resonance
CL	Caprolactone
DEG	Dietyleneglycol
DBTDL	Di-*n*-butyltin dilaurate
DEG	Diethylene glycol
DL	Dioxane lignin
DMSO	Dimethylsulfoxide
DMA	Dynamic mechanical analysis
DMF	Dimethylformamide
DSC	Differential scanning calorimetry
DTA	Differential thermal analysis
E	Compression modulus
E_a	Apparent activation energy
ESR	Electron spin resonance
FTIR	Fourier transform infrared spectrometry
GPC	Gel permeation chromatography
HL	Hydrolysis lignin

HLPCL	HL-based polycaprolactone
IR	Infrared spectrometry
KL	Kraft lignin
KLD	KL polyol derived from DEG
KLT	KL polyol derived from TEG
KLP	KL polyol derived from PEG
KLDPU	PU derived from KLD
KLTPU	PU derived from KLT
KLPPU	PU derived from KPP
KMHS	Kuhn-Mark-Houwik-Sakurada
LALLS	Static law angle laser light scattering
LCC	Lignin carbohydrate complex
LS	Sodium lignosulfonate
LSD	LS polyol derived from DEG
LSPCL	LS-based polycaprolactone
LST	LS polyol derived from TEG
LSP	LS polyol derived from PEG
LSDPU	PU derived from LS D
LSTPU	PU derived from LS T
LSPPU	PU derived from LS P
MDI	Poly(phenylene methylene) polyisocyanate
ML	Molasses
MLD	ML polyol derived from DEG
M_n	Number average molecular mass
M_w	Weight average molecular mass
MS	Mass spectrometry
MWL	Milled wood lignin
n	Number of repeating unit of oxyethylene chain
NMR	Nuclear magnetic resonance spectrometry
PCL	Polycaprolactone
PEG	Polyethylene glycol
PU	Polyurethane
SEC	Size exclusion chromatography
SEM	Scanning electron microscopy
T_1	Longitudinal relaxation time
T_2	Transverse relaxation time
TEG	Triethylene glycol
TG	Thermogravimetry
TG/FTIR	Simultaneous measurement of TG and FTIR
TG/MS	Simultaneous measurement of TG and MS
THF	Tetrahydrofuran
TMA	Thermomechanical analysis
T_c	Crystallization temperature
T_d	Thermal decomposition temperature
T_g	Glass transition temperature

T_m	Melting temperature
T_s	Softening temperature
UV	Ultra violet spectrometry
VPO	Vapor pressure osmometry
W_c	Water content
W_{nf}	Non-freezing water content
ΔC_p	Heat capacity difference at T_g
ΔH_m	Melting enthalpy

1 Introduction to Lignin

1.1 Historical Background of Lignin in Polymer Science

Polymeric components in plants, such as cellulose, hemicelluloses, and lignin interpenetrate with each other to form complex higher order structures in living plant organs in the presence of an excess amount of water in nature. The history of lignin science stretches over a period of one hundred years and many scientists had made efforts effectively to exclude the lignin from wood in order to extract cellulose in the pulping process. Lignin has been considered as an unwelcome by-product and attempts have been made to cultivate wood species having a small amount of lignin content using new biotechnology. Thus, synthetic pathways of lignin in living plant organs and also enzymatic or chemical degradation processes have been investigated.

In "Lignins", edited by Sarkanen and Ludwig [1], it is stated that the word lignin is derived from the Latin word *lignum* meaning wood. The amount of lignin in plants varies widely. In the case of wood, the amount of lignin ranges from ca. 12% to 39%, when the amount is determined according to Klason lignin analysis which is dependent on the hydrolysis and solubilization of the carbohydrate component of the lignified material [2].

The above-method consists of two steps. Lignified material is treated with 72% sulfuric acid (cooled prior to use at 10–15°C) at 20°C for a certain time and then followed by dilution of the acid to 3.0%, boiling to complete hydrolysis. The lignin is isolated as an acid-insoluble material. The content of acid-insoluble Klason lignin varies from ca. 29% to ca 39% in soft woods and from ca. 16% to ca. 22%. [2].

The chemical structures of lignin have been investigated in detail by chemical and spectroscopic methods [1–3]. Lignin is usually considered as a polyphenolic material having an amorphous structure, which arises from an enzyme-initiated dehydrogenative polymerization of *p*-coumaryl, coniferyl and sinapyl alcohols. The basic lignin structure is classified into only two components; one is the aromatic part and the other is the C3 chain. The only usable reaction site in lignin is the OH group, which is the case for both phenolic and alcoholic hydroxyl groups. Lignin consists of 4-hydroxyphenyl (**1**), guaiacyl (**2**), and syringyl (**3**) structures connected

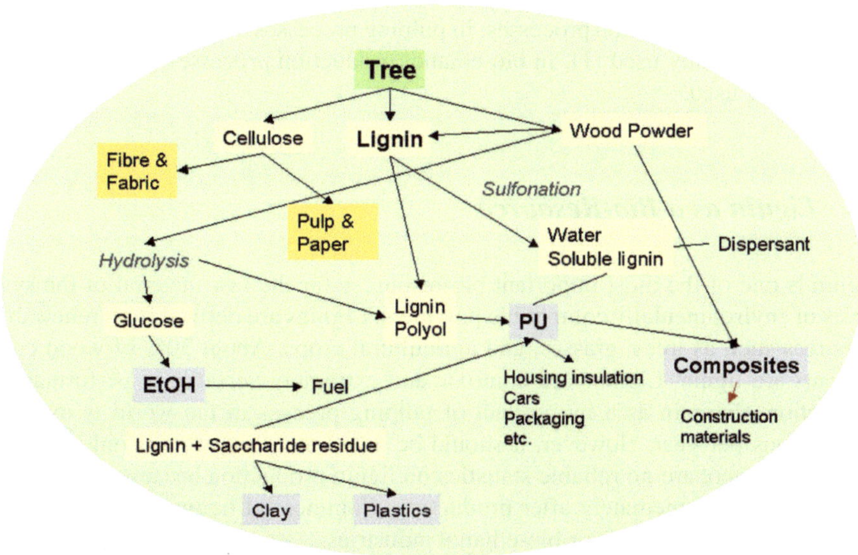

Fig. 1 Total utilization of bio-resources from wood

with carbon atoms in phenylpropanoid units, as illustrated in Fig. 1. It is essential to consider how to use the above basic lignin structures in synthetic pathways when lignin utilization for industrial materials is achieved.

The physical chemical nature of lignin as a representative biopolymer has been neglected for a long time. Goring first shed light on the polymeric nature of lignin in early 1960 [4, 5]. He measured the intrinsic viscosity of lignins and concluded that conformation of lignins is between an Einstein sphere and a non-free draining random coil in a solvent [6]. He also showed that the lignosulfonate is in the form of spherical particles of a wide range of sizes according to the electron micrograph of high molecular weight fraction of sodium lignosulfonate (LS) [6]. He also reported that the softening temperature (T_s) of lignins varied from 127 to 193°C in dry state and from 72 to 128°C in moist lignins [6].

1.2 Varieties of Lignin

Isolated lignins are usually categorized according to isolation procedures. Two approaches are generally used for the isolation of lignins from lignocellulosic materials such as wood. Acidolysis methods are based on hydrolysis of polysaccharide parts of lignocellulosics. Extraction of lignin after ball milling or enzymatic treatments of lignocellulosics gives milled wood lignin (MWL). Representative methods are introduced in various books, for example [1–3].

However, for practical approaches of effective utilization of lignins, it is important to pay attention to industrial lignins such as by-products of pulping processes and bio-ethanol production processes. In pulping processes, kraft and sulfite pulping methods are usually used [1]. In bio-ethanol production processes, acid hydrolysis is commonly used.

1.3 Lignin as a Bio-Resource

Lignin is one of the most important bio-resources for the raw material of the synthesis of environmentally compatible polymers. Lignins are derived from renewable resources such as trees, grasses, and agricultural crops. About 30% of wood constituents are lignin. Lignins are nontoxic and extremely versatile in performance. Production of lignin as a by-product of pulping process in the world is over 30 million tons per year. However, it should be noted that this value is only an estimate, since there are no reliable statistics on lignin production because it is mostly burnt as a fuel immediately after production. Commercial lignins are obtained as by-products of the pulping or bio-ethanol industries.

The higher-order structure of lignin, which consists of phenylpropane units, is fundamentally amorphous. Basically, three major structures of lignin, 4-hydroxyphenyl (**1**), guaiacyl (**2**), and syringyl (**3**) structures (shown in Scheme 1) are conjugated to produce a three-dimensional lignin polymer in the process of radical-based lignin biosynthesis. For the above reason, lignin does not have a regular structure like cellulose, but is a physically and chemically heterogeneous material, although the exact chemical structure is unknown. Since each lignin molecule has more than two hydroxyl groups, lignin-based polyurethane derivatives, polycaprolactones→PCL (PCL) derivatives and epoxy resins are obtainable by using the hydroxyl group as the reaction site.

Figure 1 shows a possible utilization plan for bio-resources from wood. Lignin is obtained as the residue of pulp and paper industry and bio-ethanol production. Even wood powder can be directly used as fillers for lignin-based polyurethane composites. It should be said that for effective and cost-saving utilization it is essential to use industrial lignin. Reasonable amounts of saccharides and polysaccharides are contaminated with lignin during the process of taking from plant-based industries.

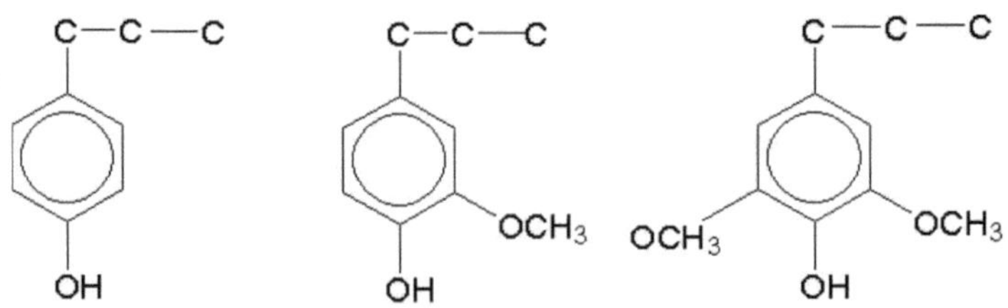

(1) (2) (3)

Scheme 1 Three important structures of lignin. 4-hydroxyphenyl (1), guaiacyl (2), and syringyl (3) structures

2 Molecular Structure and Higher Order Structure

2.1 Specific Chemical Structure

Lignin distribution in plant organs has been quantified using ultraviolet light (UV) microscopy. Using interference microscopy and confocal laser scanning microscopy, weakly or strongly lignified portion in plant organs can be visualized [7]. In plant cell walls, lignin is polymerized in a matrix of concentrated polysaccharide gel surrounded by water molecules in dynamic conditions. Synthesis using model compounds suggested that construction of the lignin polymer is affected by the pre-existing polysaccharide gel in the cell wall [8]. Thus, the structure of lignin is highly affected by environmental factors during biosynthesis. The fundamental unit structure of lignin is now well-identified owing to the research work of lignin chemists [1], while the polymeric structure of lignin has not been fully elucidated. Statistical methods in which the higher order structure is constructed by linking each possible unit compound using a computational means had received considerable attention [9–11]; however, this method has certain limitations since the constructed structure is an assumed model.

It is widely accepted that lignin is not a constitutionally defined compound, but that it is a physically and chemically heterogeneous material consisting of representative phenylpropane structures shown in Scheme 1. The phenylpropane units conjugate variously in the biosynthesis process and form a three dimensional polymer which does not have an ordered and regular macromolecular structure. The structural models of lignins have been illustrated by many authors, for example [12–14].

Generally speaking, the structure of lignin is heterogeneous. Various combinations of side chain structures and *p*-hydroxyphenyl (**1**), guaiacyl (**2**), and syringyl (**3**) structures (see Scheme 1) increase the diversity of lignin structures. The detailed structures of lignins are academically important; however, the elucidation of the complicated structural linkages of each phenylpropane unit of lignin does not lead directly to the effective utilization of lignins. It may be appropriate to consider that the diversity of structural combinations is unlimited because of various factors influencing the biosynthetic process of lignin in plants.

Accordingly, it is worth considering the structural moiety of various lignins suggested by Balakshin [14]. It is also necessary to realize the importance of the amount of different functional groups in natural and industrial lignins [14]. According to current understanding of lignin, almost all lignins have linkages to carbohydrates [15]. This indicates that contaminants of fragments of carbohydrate are unavoidable in isolated lignins because of the presence of lignin-carbohydrate complex (LCC). Accordingly, in the utilization of isolated lignins, it should not be too sensitive to carbohydrates contaminants in technical lignins obtained as by products of modern industries such as pulping and bio-ethanol production. Modern synthetic methods may solve the problems of the presence of carbohydrate contaminants (ref. Chaps. 5 and 6).

2.2 Molecular Mass and Molecular Mass Distribution

Molecular mass of isolated lignin has been observed by usual techniques, such as viscometry [16], size exclusion chromatography (SEC), gel permeation chromatography (GPC) [17–20], light scattering [21, 22], VPO [22–24] and ultracentrifugation [25]. When the solution properties of isolated lignin are measured, the major problem is the solubility in organic solvents. As described in Sect. 1, lignin molecules are slightly cross-linked and this suggests that the insoluble portion is necessarily obtained. It is thought that unavoidable uncertainty comes from the fact that the molecular mass of soluble portion of lignin is comparably smaller than that of really isolated sample. This point becomes more complicated since solution properties of lignin are thought to depend not only on plant species but also on the isolation process [26].

The exponential factors of Kuhn-Mark-Houwink-Sakurada (KMHS) equation for kraft lignin (KL) are 0.11, 0.13, and 0.23 in dimethylformamide (DMF) at 45.2°C, in DMF at 77.7°C, and in 0.5 N sodium hydroxide at 30.2°C, respectively [16]. The fact that KMHS exponential factors of KL in DMF are small indicates that the molecular mass of lignin scarcely affects the reduced viscosity. This suggests that the lignin molecules in DMF have a compact spherical structure and approach the limit of an Einstein sphere, a constant-density sphere. The above results had also been reported by the research group of Goring [4].

GPC results of fractionated lignin are also obtained and the molecular mass is in a range from 2.0×10^3 to 7.5×10^3 [17]. The effects of processing on molecular mass and molecular mass distribution of spruce and pine lignin were measured by GPC [27]. By fractionation, it is reported that average molecular mass (M_w) ranges from 2.7×10^3 to 1.1×10^4 and molecular mass distribution ranges from 1.8 to 2.4 for organosolv lignins.

Lignosulfonate is mainly obtained as by-products of the sulfite pulping process, and chemical structures of LS have been illustrated by Glennie [28]. LS is an amphiphilic polymer and is water soluble. Goring reported that the polyelectrolyte LS forms of spherical particles of a wide range of sizes and particles of the high molecular weight fraction of LS were observed by electron micrograph [29]. Results of detailed characterization of lignosulfonate were recently reported by Lebo [30]. According to the above article [30], M_w is $6.4 \times 10^4\ (\mathrm{gmol}^{-1})$, M_w/M_n 8.8, where M_n is number average molecular mass, and fractionated samples shows from M_w, 4.6×10^3 to M_w, $4.0 \times 10^5\ (\mathrm{g\ mol}^{-1})$. These results indicate that molecular weight distribution of lignosulfonate is large and molecular fractions of lignosulfonate are widely spread.

By introducing acetyl or hydroxypropyl group to lignin, lignin becomes soluble in tetrahydrofuran (THF) or other organic solvents, which are commonly used to investigate the solution properties of polymers. GPC was carried out using various types of acetylated semi-commercial lignin [31]. The exponential factors of KMHS equation for acetylated KL, organosolv, and steam explosion lignin is in the range of 0.17–0.35. KL was fractionated and acetylated. Molecular mass of each fraction was measured by GPC [18]. Molecular mass distribution increased with increasing

M_W. Molecular mass and distribution of hydroxypropylated lignins obtained from various trees were also measured using GPC, static law angle laser light scattering, and VPO [32]. Molecular mass distribution of lignin in situ was observed by SEC. By using the difference of refractive index and UV light between cellulosic components and lignin, SEC profile of lignin in wood can be magnified. M_W of lignin was in a range ofca. 1.9–2.4 $\times 10^4$ according to various bleaching processes [33].

The above papers suggest that the molecular mass of isolated lignin is in the range of 10^3–10^5 depending on plant species, processing method, and also measuring method. Molecular mass distribution depends on plant species, isolation methods, and measuring conditions. Lignin molecules in solution are approximately spherical particles and slightly solvated with solvent.

2.3 *Morphology by Atomic Force Microscopy*

Atomic force microscopy (AFM) has been used in order to observe the supermolecular structure of LS by using aqueous solutions as samples, since these samples were easy to spread on a freshly cleaved mica surface. Figure 2a, b shows a representative AFM image of LS with concentration 10^{-3} wt % extended on mica surface. LS was dissolved in purified water and a solution with 10 $\mu g\,mL^{-1}$ was prepared. The above solution was diluted according to measurement conditions. The solution was dropped onto the newly cleaved mica surface thus obtained. A Seiko Instruments scanning probe microscope, SPA400, was used. Frequency was 1 Hz and measurements were carried out by tapping mode. Figure 2a, b show that lignin has a complicated network structure which is cross-linked.

The value of measured width of the sample was calibrated taking into consideration the geometrical shape of the needle, i.e. the locus of the cantilever tip is larger than the real size of the particle, since the size of the tip (radius = 10 nm) is larger than that of a sample. In this study, the cross section of the sample was assumed to be a sphere (radius = a) and real width (2a) was calibrated from apparent

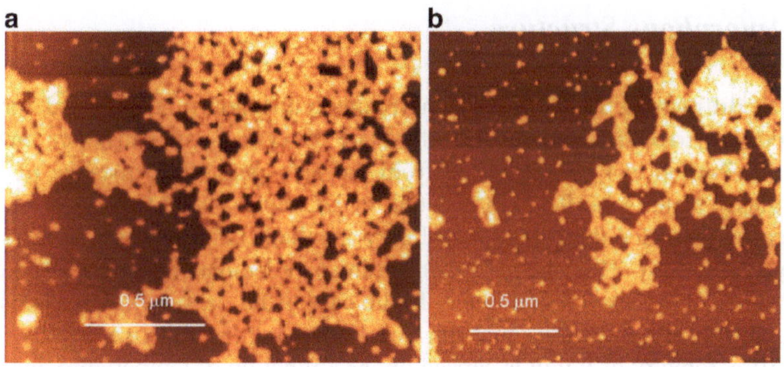

Fig. 2 AFM images of LS with concentration of 10^{-3} wt %

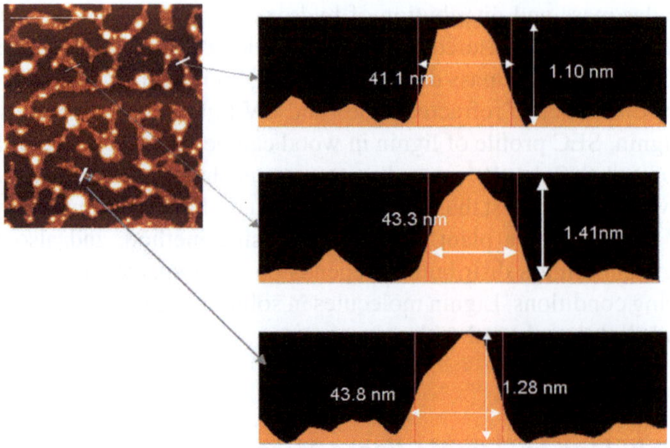

Fig. 3 Molecular sizes of LS measured by AFM. concentration = 10^{-3} wt%

width (W) by $W = 4\sqrt{10a}$. The average calibrated width was in a range from 11.7 nm, depending on the portion of bundles and the standard deviation was 3.6 nm. Size distributions of representative samples are shown in Fig. 3. The height is mostly distributed from 1.1 to 1.4 nm. The evenness of mica surface was also analyzed. The variation of the height ranges from 0.05 to 0.10 nm. Thus, the evenness of the mica surface must be taken into consideration when the standard deviation of the height of bundle-like structure is noted.

The above facts indicate that LS molecules form a cross-linked network structure. However, the shape of LS molecules changes with the concentration of aqueous solution. This strongly suggests that lignin molecules form a slightly different network structure depending on a change in the state of aggregation. The results of AFM observation accord well with the physical properties of lignin in solid state.

2.4 Amorphous Structure

By x-ray diffractometry, a typical amorphous halo pattern is observed for various types of isolated lignin in the solid state. Inter-molecular distance having a broad distribution can be calculated from the halo peak. The distance of MWL in the glassy state is ca. 0.42 nm and that of DL corresponds to ca 0.43 nm [34]. The peak is broader than that of ordinary synthetic amorphous polymers such as polystyrene [35]. This suggests that the inter-molecular- distance of molecular chains is distributed in a broad range. In the case of DL, a sub-peak was observed at around 0.98 nm, indicating that intra-molecular regularity exists locally. When x-ray diffractometry of lignin is carried out as a function of temperature, inter- and intra-molecular distance increase at a characteristic temperature which corresponds to glass transition temperature → Tg. Glass transition of various kinds of lignin is described in Chap. 3.

3 Physical Properties in Solid State

3.1 Glass Transition

Molecular motion of lignin is characterized by phenylpropane units in the molecular chain; bulky side chain molecules and slight cross-linking establish intra- and intermolecular chains. Rigid groups in the main chain and cross-linking restrict molecular motion and have the effect of increasing the glass transition temperature (T_g). In contrast, bulky side chains enhance molecular mobility of lignin through the local mode relaxation. As described in Chap. 2, the chemical structure of lignin depends on plant species and also the method of isolation. Reflecting the variety of chemical components and complex higher-order structure, the temperature range of glass transition is broader than that of synthetic amorphous polymers, such as polystyrene with the same molecular mass. When the ionic group is introduced in the molecular chain, molecular motion of polyelectrolyte lignin is markedly restricted.

Among various techniques to measure glass transition behavior of amorphous polymers, differential scanning calorimetry (DSC) is widely used, since it has several advantages, such as small sample size and ease of control of the thermal prehistory [36]. Figure 4 shows representative DSC heating curves of various kinds of industrial lignin, such as LS, KL and hydrolysis lignin → HL (HL) obtained from different industrial plants. Heating rate was 10°C min^{-1}, sample mass was ca. 5 mg. Prethermal history of the sample was erased by annealing at 180°C and then quenched to a temperature lower than −100°C. Baseline shifts to endothermic direction are observed and T_g was defined as temperatures indicated by arrow. T_g of LS and KL is observed at ca. 100°C; however, T_g of HL samples obtained from various industrial plants are observed in a temperature range of 70–85°C.

Amorphous polymers show enthalpy relaxation when glassification slowly takes place. By annealing at a temperature at around T_g, the same effect can be obtained as slow cooling. Figure 5 shows representative DSC heating curves of HL (New Zealand) annealed at 70, 80, and 90°C for 4 h, cooled and then heated at the rate of 10°C min^{-1}. By annealing, enthalpy relaxation is identified as an endothermic shoulder peak in DSC curves as indicated by arrows. As shown in the figure, with increasing annealing temperature, the endothermic shoulder peak shifts to the high temperature side. DSC curves shown in Fig. 4 indicate that lignin is a typical amorphous polymer having a broad spectrum of molecular higher order structure which coaggregates by annealing. Heat capacity difference at T_g of HL-1 is 0.40–0.40 j $\mathrm{g}^{-1}\,\mathrm{K}^{-1}$, indicating that molecular motion is not markedly restricted compared with ΔC_p values of typical amorphous polymers [37]. By annealing, ΔC_p values decreased with increasing annealing time, suggesting that enthalpy of glassy

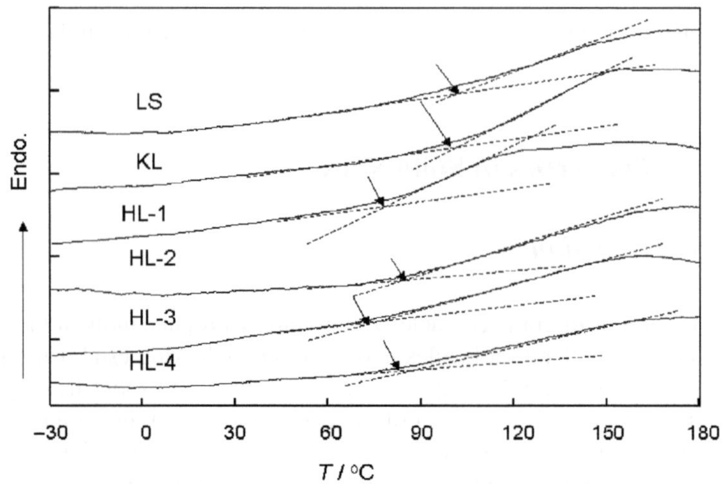

Fig. 4 Representative DSC curves of LS, KL and HLs obtained from various industrial plants. Heating rate = 10°C min^{-1}. *LS, KL, HL-1* New Zealand; *HL-2* Russia (Archangelsk plant); *HL-3* Russia (Lobinsky plant); *HL-4* Russia (Kirovsky plant)

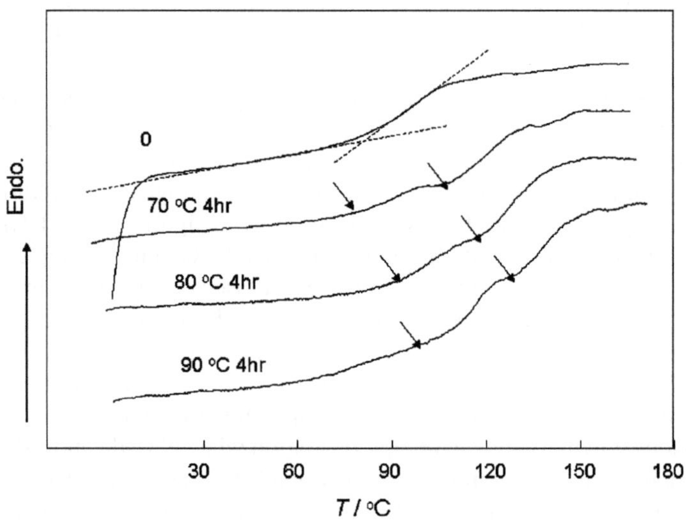

Fig. 5 DSC heating curves of HL-1 annealed at various temperatures and then heated at 10°C min^{-1}. *Numerals* in the figure show annealing temperature. Annealing time = 4 h

state of lignin stabilizes by annealing. The above results suggests that cross linking of lignin molecules are not as dense as suggested in the model structures shown in previous papers [9–11].

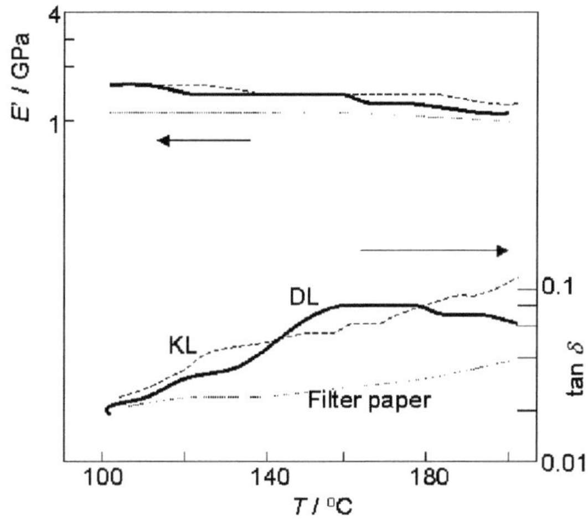

Fig. 6 Temperature dependence of the dynamic viscoelasticity of DL-filter paper and KL-filter paper composites

Figure 6 shows dynamic viscoelasticity of DL and KL. In order to measure the dynamic viscoelasticity, films or sheets are required. Since lignin samples were in powder form, THF solution of DL or KL was prepared and strips of filter paper were immersed in the above solution, and then dried completely. Filter paper without lignin shows no transition in a temperature range of 100–200°C. Decrease of dynamic modulus (*E*') was observed at around 160°C, tan δ peak is clearly observed for DL. A broad shoulder in tan δ curve is also observed for KL at around 120°C [38].

Glass transition behavior of lignin measured by various experimental techniques, i.e. dilatometry [5], DSC [17, 39–44], viscoelastic measurements [38, 45, 46], dynamic mechanical analysis (DMA) [47], thermomechanical analysis (TMA) [48], solid state nuclear magnetic resonance spectrometry (NMR) [49], spin probe method using electron spin resonance spectroscopy (ESR) (temperature dependency of number of radicals which can be found in isolated lignin) [50], temperature dependency of characteristic absorption bands measured by infrared spectroscopy (IR) [51], and temperature dependency of x-ray diffractogram [34] have been reported. Table 1 shows T_g values of isolated lignin and chemically modified lignin measured by various methods. Glass transition is a relaxation phenomenon and thus T_g value depends on time factors, such as frequency and heating rate. At the same time, rate of glassification also affects the heat capacities of the glassy state due to enthalpy relaxation [35]. Subsequently T_g value depends on the thermal history of each lignin sample. Thus, T_g values shown in Table 1 have certain errors. T_g values of several synthetic amorphous polymers having the same structure shown in Scheme 1 in the side chain (See Sect. 1) are also shown in this table.

Table 1 Glass transition temperature of various kinds of lignin and lignin derivatives measured in dry state measured by different techniques

Sample	$T_g/°C$	Measurement method	References
Lignin in situ			
Ponderosa pine[a]	80	DMA	[48]
Lignin, isolated			
MWL	120	IR[b]	[51]
MWL(1)	170–180	TMA	[47]
MWL	ca.120	ESR[c]	[50]
DL	ca.105	dilatometry	[5]
DL	120–150	DSC	[40, 52]
DL	ca.120	Mechanical	[38]
DL	ca.130	X-ray[d]	[40]
DL	ca.130	b-NMR[e]	[49, 53]
KL	ca.130	b-NMR	[49, 53]
KL	110	DSC	[44]
LS	ca.130	DSC	[39]
LS	ca.130	b-NMR	[49, 53]
CaLS	ca.120	ESR	[50]
HL	70–85	DSC	[54]
Organosolv lignin	109	DSC	[44]
Lignin derivatives			
Hydroxypropyl KL	71	DSC	[44]
Carboxymethyl KL	129	DSC	[44]
Acetoxypropyl KL	47	DSC	[44]
Acidolysis lignin	120	DSC	[52]
Reference Polymers			
Poly(4-hydroxystyrene)	182	DSC	[55]
Poly(4-hydroxy, 3-ethoxystyrene)	142	DSC	[55]

[a]Poplar tree, 2 years and 6 month

[b]T_g was defined at a temperature where the relative optical baseline density of IR absorption band at 1,500–1,510 cm^{-1} decreased steeply

[c]T_g was defined at a temperature where the number of radicals decreased steeply

[d]T_g was defined at a temperature where inter-molecular distance increases when halo pattern of lignin was measured as a function of temperature

[e]T_g was defined at a temperature where the second moment decreased steeply

3.2 Local Mode Relaxation

Local mode relaxation of isolated lignin and its model compounds have been detected by dynamic mechanical measurement, and broad-line nuclear magnetic resonance spectroscopy (b-NMR) [49, 53], although this molecular motion has scarcely received attention in recent papers. Transition map of local mode relaxation of various kinds of polymers is found elsewhere [56]. Figure 7 shows second moment of absorption line of b-NMR of DL in powder form. When the relaxation is from the

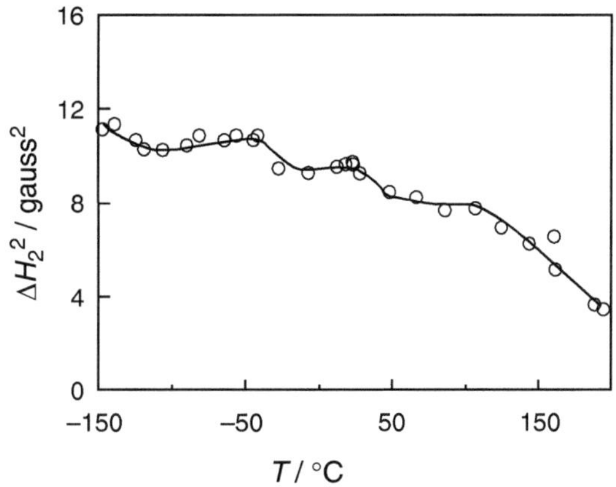

Fig. 7 Relationship between second moment of absorption line of b-NMR of DL

high temperature side to the low temperature side, two local mode relaxations, β and γ, can be observed for various types of isolated lignins, such as DL, LS, and MWL, besides α relaxation which corresponds to main chain motion (glass transition). Model polymers having basic chemical structure of lignin (see Scheme 1), such as poly(4-hydroxystyrene), and poly(3-methoxy, 4-hydroxy styrene), showed two local model relaxations at a similar temperature range of lignin. From the value of second moment and results of other synthetic polymers [49], β relaxation is attributed to the hindered motion of methyl group and γ relaxation is attributed to the hindered motion of the side chain [49].

3.3 Lignin–Water Interaction

Lignin in situ exists in an excess of water in living plant tissue in which carbohydrate polymers and lignin are believed to form complex structures having chemical linking. Accordingly, physical properties of lignin in situ have been investigated by measuring plant tissue in the presence of water or humid conditions [for example, 57,58]. This suggests that molecular motion as described in Sects. 3.1 and 3.2 is markedly affected by coexisting carbohydrate polymers and water molecules in the tissue. Molecular properties of hydrophilic polymers in the presence of water have received particular attention since the functionality of biopolymers is markedly affected by water [59]. Lignin is hydrophobic compared with carbohydrate polymers in plant tissue; however, the number of hydroxyl groups attaching to lignin is sufficient to act as a reaction site for hydrogen bond formation with water molecules. From the results obtained using polyhydroxystyrene derivatives having the same structure as shown in Scheme 1 in their side chain, the hydroxyl group attached adjacent to the methoxyl groups plays an important role [55]. In this chapter the molecular motion of isolated lignin in the presence of water is introduced. Molecular mobility of lignin is markedly enhanced by adding a small amount of water [52].

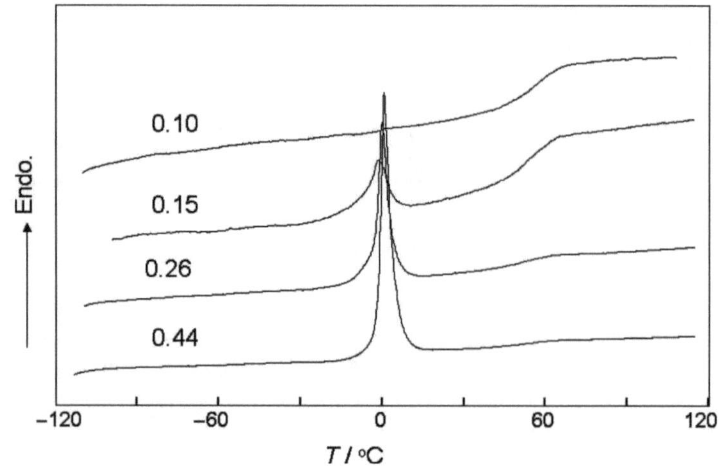

Fig. 8 DSC heating curves of HL-1 with various water contents. Heating rate = 10°C min^{-1}, in N_2 atmosphere. *Numerals* in the figure show water content

Molecular motion of water restrained by lignin molecules is described, since structural change of water molecules restrained by matrix biopolymers is observed.

Figure 8 shows DSC heating curves of HL with various water contents. Water content is defined as follow:

$$W_c = m_w / m_s. \quad (1)$$

Where m_s is mass of dry lignin sample and m_w is mass of water in the lignin-water system. T_g value of dry HL is lower than that of other isolated lignin. As shown in Fig. 8, T_g shifts to the low temperature side with increasing W_c. At the same time, heat capacity difference between glassy state and rubbery state (ΔC_p) at T_g increases, suggesting that molecular motion of lignin is enhanced in the presence of water molecules. As shown in Fig. 8, when W_c exceeds 0.15, an endothermic peak due to a melting of ice in the HL-water system is clearly seen.

Figure 9 shows the relationship between T_g and water content (= W_c) of HL-1 (from New Zealand plant) and DL. T_g value decreases markedly in the initial stage of water sorption (W_c ranges from 0 to 0.1) and then levels off. The T_g difference of HL-1 at completely dry state ($W_c = 0$) and T_g at leveling off is ca. 40°C and that of DL [52] was ca. 90°C. T_g difference between dry and wet hydrophilic model polymers whose structure has lignin basic structure is also 80–90°C [55]. The above facts suggest that a small amount of water breaks the inter-molecular hydrogen bonding of lignin and molecular motion is enhanced markedly.

Water induced softening of LS with various counter ions, such as Li, K, Na, tetramethylene ammonium ion and divalent ion Ca was investigated by torsion braid analysis in humid controlled atmosphere [45, 46]. The position of dumping (Δ) peak depends on temperature and counter ions. It was found that the relative humidity showing Δ peak decreased with increasing radius of monovalent cations [46]. DMA

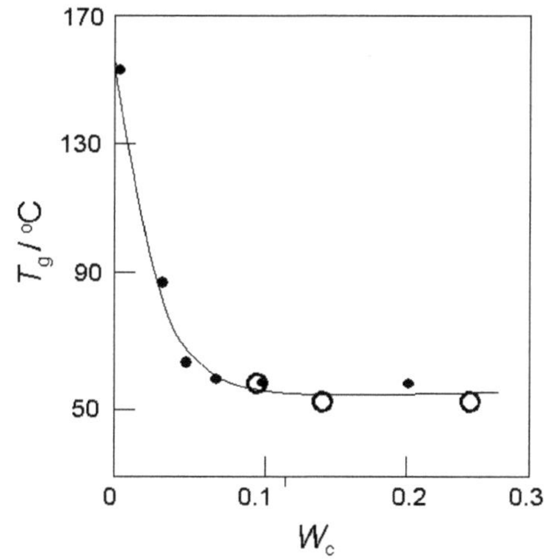

Fig. 9 T_g of HL and DL with various water contents as a function of water content. *Open circle*, HL; *closed circle*, DL

of various kinds of wood in water was carried out and T_s was observed in the range 70–95°C [58]. It was found that T_s decreased with increasing methoxyl groups in lignin. The above results accord well with molecular behavior of isolated lignin. Although T_s is an important index for polymer processing, T_s not directly accord with T_g in the cases of crystalline polymer. Accordingly, T_s of wood is not directly defined as T_g of lignin since cellulosic component in wood works as crystalline part of the whole polymer.

When hydrated hydrophilic polymers are measured by DSC, three kinds of water are identified from temperature and enthalpy of transition peaks. Hatakeyama and coworkers classified hydrated water (water in polymers) as follows: freezing (crystallizable) water which can be observed as the first order phase transition and grouped into (1) free water which crystallizes at ca. 0°C and is slightly affected by matrix polymer or nonfreezing water and (2) freezing bound water which crystallizes in the heating process at a temperature lower than 0°C strongly affected the matrix polymer. Besides freezing water, there exists (3) nonfreezing water i.e. water which is non-crystallizable due to strong molecular interaction with matrix polymer. Both freezing bound water and nonfreezing water are categorized as bound water [59, 60]

Enthalpies of transitions were calculated from the peak area of the first-order thermodynamic transition. When the two melting peaks were observed, enthalpy of each peak was calculated. In the case where two melting peaks are observed, bound water was calculated as follows:

$$W_c = W_f + W_{fb} + W_{nf}. \quad (2)$$

$$W_b = W_{fb} + W_{nf}, \quad (3)$$

where W_f is amount of free water calculated from melting enthalpy (ΔH_m) of the high temperature side melting peak, W_{fb} is freezing bound water calculated from

ΔH_m of the low temperature side melting peak, and W_b is the total amount of bound water. The amount of nonfreezing water (W_{nf}) can be calculated by equation 2, since W_c was obtained from equation 1.

As shown in Fig. 8, DSC heating curves of water restrained by HL-1 at $W_c = 0.1$, no melting peak is observed, although glass transition is clearly seen All water molecules form hydrogen bonding with the hydroxyl groups of HL. It is thought that water molecules are separately located near the hydroxyl group of lignin molecules. Thus, the number of water molecules to form ice is not sufficient in the system With $W_c = 0.15$, a broad melting peak is observed at around $-1°C$. However, it is clearly seen that melting peak starts at around $-30°C$. As described above, this kind of water is categorized as freezing bound water. From the melting enthalpy of this peak W_{fb} was calculated. In HL with W_c higher than 0.26, the melting peak became sharper and narrower suggesting that free water is observed beside freezing bound. In HL-1, separation of freezing bound and free water was not distinct. From ΔH_m of the melting peak, W_{nf} was calculated using (2). Figure 10 shows the relationships between W_{nf} and W_c. The amount of W_{nf} of HL-1 is ca. 0.2 which is not large compared with other biopolymers, for example wood cellulose 0.4 and sodium cellulose sulfate 0.43.

Molecular mobility of water molecules restrained fromLS was observed from ^1H and ^{23}Na relaxation times by NMR. The longitudinal relaxation times (T_1) were measured by the 190-τ-90 degree pulse technique and transverse relaxation times (T_2) were obtained by Meiboon-Gill variant of the Curr-Purcell method, the Hahn spin-echo and from the free induction decay following the 90-degree pulse, which were chosen according to W_c of LS. W_c of LS for ^1H measurements was in the range from 0.6–2.2 and the temperature range was from 50–120°C. Figure 11a,b shows T_1 and T_2 values obtained by ^1H NMR at representative temperatures 20 and $-20°C$. W_{nf} of LS used in this study was 0.58 [61]. Since LS is water soluble, the W_{nf} value of LS is much larger than that of HL or DL. Thus, LS samples measured by NMR contains nonfreezing and free water since the amount of freezing bound

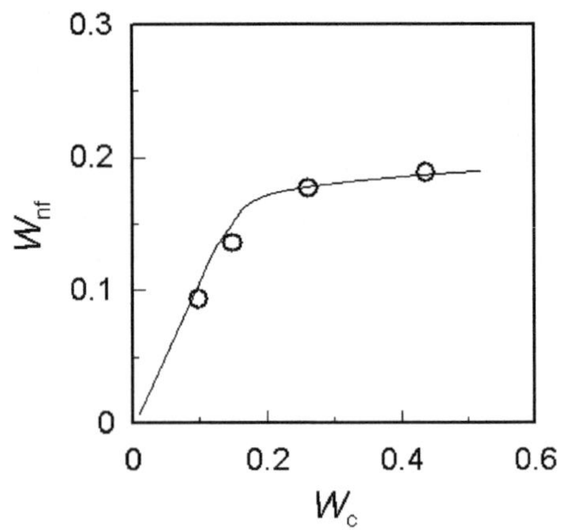

Fig. 10 Amount of W_{nf} restrained by HL as a function of water content

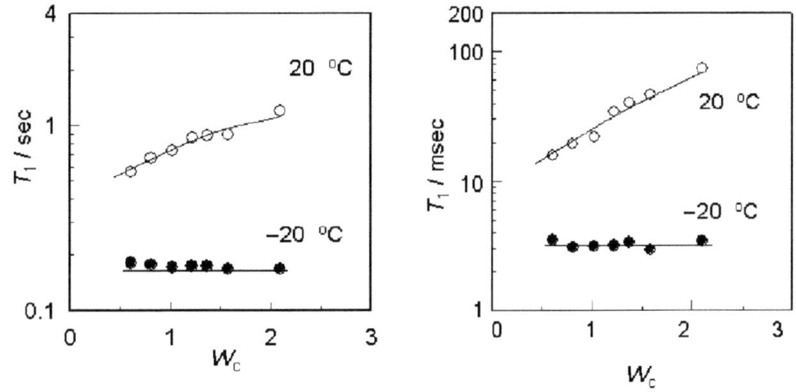

Fig. 11 ^{1}H longitudinal relaxation time (T_1) (**a**) and transverse relaxation time (T_2) (**b**) of water restrained by LS as a function of water content (W_c)

water was negligible. At 20°C, both T_1 and T_2 values increase with increasing W_c; in contrast values at −20°C remained almost constant. When T_1 values were plotted against reciprocal absolute temperature, it was clearly seen that the extreme narrowing condition was fulfilled at around −20 to −25°C. This indicates that the T_1 value at −25°C is the lowest value. At the same temperature, T_2 values showed a sudden decrease. This indicates that molecular motion ceases at −25°C since the T_2 values represent an average motion of water molecules. Figure 11 indicates that the amount of mobile fraction of water increases with increasing W_c. When the results of Fig. 11 are compared with those of thermal data having the same thermal history, it is clear that the molecular mobility of water molecules was restrained by LS increases with increasing W_f. By freezing of the water (crystallization of free water), molecular motion ceases and relaxation time levels off. From thermal data it is known that nonfreezing water freezes at T_g, that is lower than −20°C. This explains why T_1 and T_2 values of water with LS at $W_c = 0.60$, that mostly consist of W_{nf}, are slightly higher than those of LS systems.

By Woessner's equations, the correlation time of bound water is calculated using the T_1, T_2 values and angular resonance frequency based on several assumptions [62, 63]. The correlation times (τ_c) of bound water (nonfreezing water in this case) of LS calculated by assuming one average value are shown in Fig. 12. When molecular motion of all free water ceases, τ_c values remain constant. With increasing reciprocal absolute temperature τ_c increases linearly regardless of W_c. Thus, apparent activation energy E_a can be calculated. E_a of bound water of LS was $24\,\mathrm{kJ\,mol^{-1}}$ [61].

Molecular mobility of Na ion in LS was also studied by ^{23}Na NMR in a W_c range from 0.5 to 1.5. Figure 13 is a three-dimensional figure between ^{23}Na T_1, and W_c and shows reciprocal absolute temperature of LS. When the temperature decreased below −20°C, the $180-\tau-90$ pulse method could not be applied due to marked line broadening. This indicates that free water molecules freeze and form ice and the mobility of Na ions is restricted. The gradient of T_1 values against reciprocal

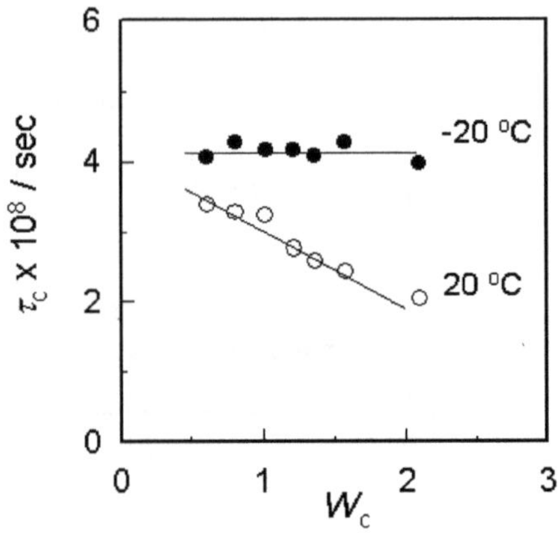

Fig. 12 Relationships between correlation time (τ_c) at 20 and −20°C and water content of water restrained by LS

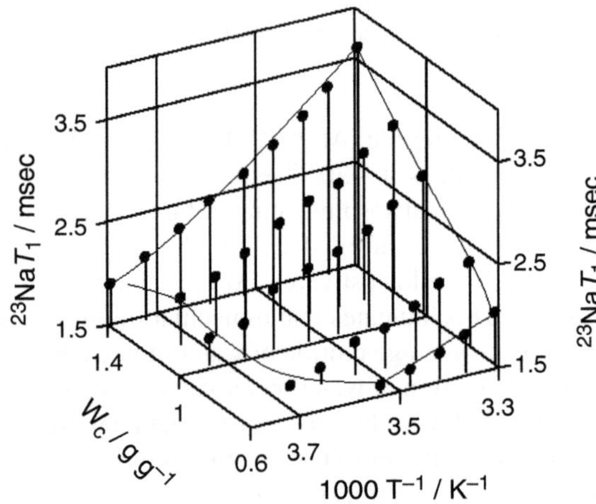

Fig. 13 Relationships between, ^{23}Na T_1, W_c and reciprocal absolute temperature of LS

absolute temperature is almost the same as shown in Fig. 13.Thus, E_a of relaxation process was calculated. The calculated value was 12 kJ mol^{-1} [61].

4 Chemical and Thermal Decomposition

4.1 Chemical Decomposition

Chemical degradation of lignin is a field that has been well researched. Although a number of methods have been proposed, several methods are generally known. The following methods can be considered as major chemical degradation methods of lignin: acidolysis (hydrolysis), nitrobenzene oxidation, permanganate oxidation, hydrogenolysis, ozonization, and thermolysis.

4.1.1 Acidolysis (Hydrolysis)

Acidolysis was introduced in lignin chemistry as a procedure for the isolation of lignin, mainly from woody materials [64]. This method is one of the most widely used methods to isolate lignin easily from wood, since the method consists of the acid treatment of lignin in a water-dioxane solution. This method is also used for the cleavage of ether linkage. Some examples of representative chemical structures obtained by acidolysis are shown in Scheme 2 [65–68].

Acidolysis is close related to hydrolysis that is used for bio-ethanol production. The decomposition mechanisms and hydrolysis products of lignocellulosics under various conditions in hot-compressed water were discussed in order to establish preferable conditions among hydrolysis behavior in hot-compressed water, acid, alkaline, and enzymatic hydrolysis [69]. Hydrolysis lignin obtained as a byproduct of bio-ethanol industries can be used as the raw material for the production of lignin-based new environmentally compatible polymers (see Chaps. 5 and 6).

4.1.2 Nitrobenzene Oxidation

It is known that nitrobenzene oxidation of lignins was originally introduced by Freudenberg [70] in order to investigate lignin structures. It was reported that nitrobenzene oxidation of softwood lignin gave vanillin (**9**) as a major product and that hardwood lignins gave syringaldehyde (**17**) as oxidation products [71, 72]. Higuchi reported that grasses and their lignins gave 4-hydroxybenzaldehyde (**25**), vanillin and syringaldehyde (**17**) as oxidation products by nitrobenzene [73]. A difference in oxidation products was recognized according to the portion of wood [74].

4.1.3 Permanganate Oxidation

Permanganate oxidation was introduced for the analysis of lignin structure by Freudenberg [75]. Aromatic carboxylic acids such as 4-hydroxyphenolic acid (**25**), vanillic acid (**9**), and syringic acid (**18**) were found by permanganate oxidation of lignins [76]. More research work was carried out on oxidation of residual lignins in sulfite and kraft pulps in order to elucidate lignin structures [77–82].

4.1.4 Hydrogenolysis

Sakakibara reported in detail on hydrogenolysis of lignins [83]. He mentioned that hydrogenation and hydrogenolysis are different processes and defined them as follows. Hydrogenation is the addition of hydrogen (H_2) to a multiple bond. Hydrogenolysis is a process in which cleavage of a carbon-carbon or carbon-hetero atom single bond is accomplished by reaction with hydrogen. Examples of monomeric aromatic products in the hydrogenolysis of lignins are as

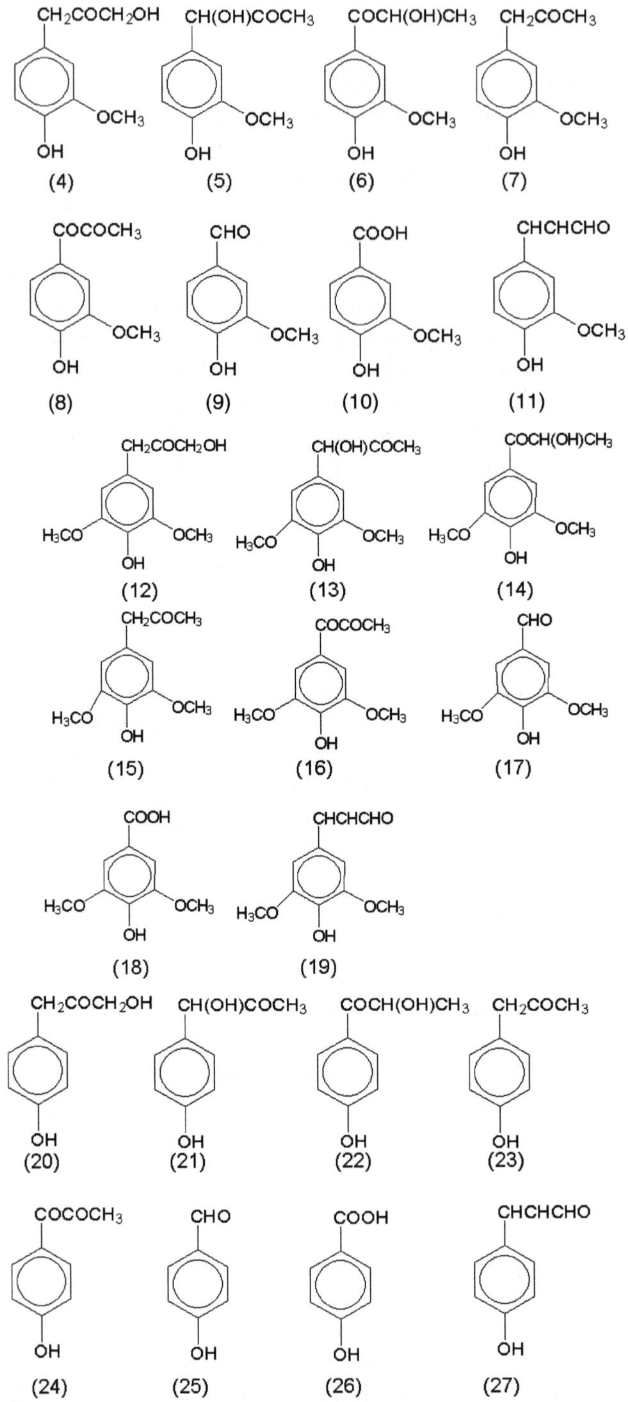

Scheme 2 Examples of chemical structures of acidolysis lignin

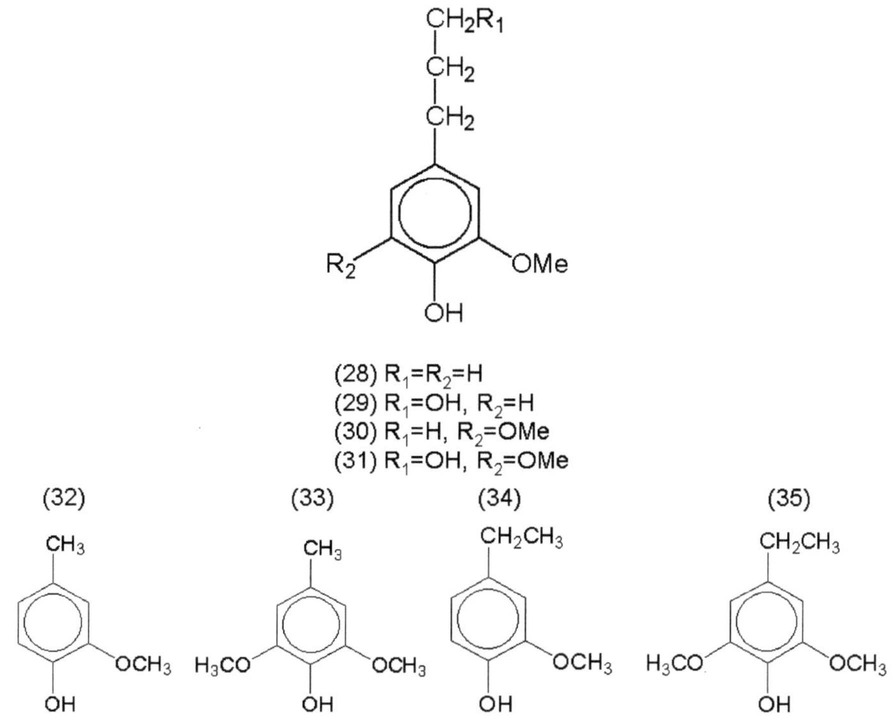

Scheme 3 Examples of chemical structures of hydrogenolysis of lignin

follows: 4-n-propylguaiacol (**28**), dihydroconiferyl alcohol (**29**), 4-n-propylsyringol (**30**), dihydrosinapyl alcohol (**31**), 4-methylguaiacol (**32**), 4-methylsyringol (**33**), 4-ethylguaiacol (**34**) and 4-ethylsyringol (**35**) (see Scheme 3).

4.1.5 Ozonization

Ozone is used for the cleavage of carbon-carbon double and triple bonds. Accordingly, ozone is effective in the degradation of lignin and can be used for bleaching of pulps. Various lignin model compounds were used to study the mechanism of lignin degradation. Examples of model compounds are mostly monomeric and dimeric lignin model compounds [84]. For example, ozonization of lignin model compounds was carried out using vanillyl alcohol (**36**) and veratoryl alcohols [85]. As the degradation product of the above model compounds (**36**), δ-lactone of the monomethyl ester of β-hydroxymethyl muconic acid (**39**) was obtained. Accordingly, it was considered that demethylation of veratoryl alcohol occurred at the 4-position of aromatic ring followed by aromatic ring opening between the carbon atoms of the 3- and 4-positions of aromatic ring (see Scheme 4). In the case of the ozonization of the same model compounds under basic conditions, δ-lactone of β-hydroxymethyl

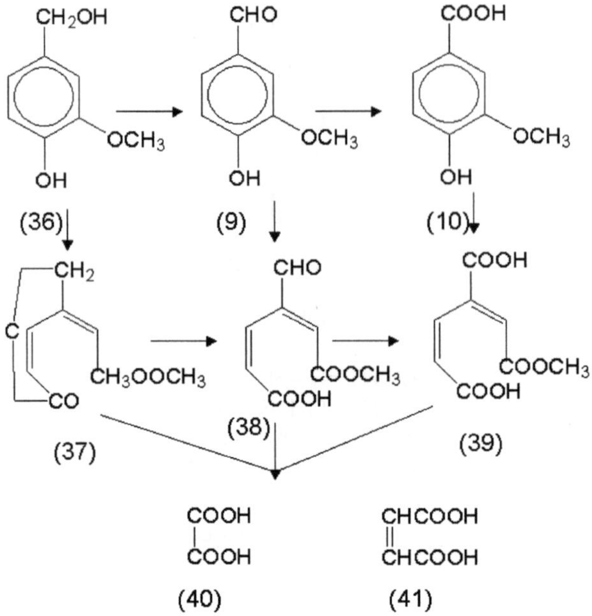

Scheme 4 Decomposition of lignin model compound by ozonization

muconic acid was obtained as the degradation product. Accordingly, it was considered that demethylation occurred at the 3-position of aromatic ring. It was found that oxidation of the side chain was much less than aromatic ring opening.

Erythro/threo ratio for lignin side chains was intensively studied by Tsutsumi et al. [86]. It is known that the erythro/threo ratio affects the rate of delignification during kraft pulping process, since etherified erythro β-O-4 structures hydrolyzed more rapidly than threo structures [87]. The kinetics of degradation [88] in aqueous acetic acid, the influence of a catalytic amount of manganese cyclam derivative on ozonization reaction of lignin model compounds [89], and also comparison of ozonization in methanol/water and water [90] have been reported. Based on the molecular mass and the molecular mass distributions, a random stepwise mechanism was suggested as the mode of lignin degradation by ozone [91].

4.1.6 Thermolysis

Thermolysis of lignin was carried out in ethylene glycol and diethylene glycol (DEG) at 300 and 350°C and the obtained oil was characterized [92]. The thermolysis of organosolve lignin in supercritical water or in supercritical methanol has been investigated at a temperature range of 350–425°C under pressure of 10–40 MPa [93]. Oils with high concentration of hydroxyl groups were obtained by reactions in supercritical water. In contrast, oils having high aromaticity were obtained by reactions in supercritical methanol. Thermolysis of lignin and petroleum waste was examined in order to develop a new usage of lignin [94]. Simulation of lignin thermolysis under various conditions has been carried using a mathematical model. A model compound was reacted in supercritical water and change in the amount decomposed by different reaction conditions was reported [95]. Modeling entailed a statistical interpretation of lignin structure [96].

4.2 Thermal Decomposition

Thermal decomposition of lignin has mainly been investigated for three different purposes: (1) pyrolytic method is appropriate in order to investigate the structure of lignin due to rapid analysis using small amounts of samples; (2) reaction process of thermal degradation is necessarily required in order to produce carbon from lignin or charcoal from wood; (3) characterization is necessary in order to utilize decomposed materials as source of chemicals. Among various techniques in order to fulfill the above purposes, pyrolysis at isothermal condition, thermogravimetry (TG), simultaneous measurements of TG/Fourier transform infrared spectrometry (TG/FTIR), and TG/mass spectrometry (TG/MS) are used in both dynamic and static conditions. Various kinds of lignin, for example KL [97], solvolysis lignin [42], DL [41, 98], MWL [99], MWL [47] have been investigated by TG, Alcell lignin [100], KL [101], KL [56] by TG/FTIR, KL [102] by pyrolysis measurement, and KL by pyroprobe-gas chromatography/MS [103]. Recently, various bioresources for further conversion to produce biofuels have attracted attention [103, 104]

- Figure 14 shows TG curves and derivative TG curves of HL which were obtained as by-products of bio-ethanol production in New Zealand (HL-1) and Russia (Archangelsk (HL-2), Lobinsky (HL-3) and Kirovsky Plants (HL-4)), KL $(M_n = 1 \times 10^3,\ M_w/M_n = 2.2)$ obtained as a by-product of kraft pulping process (Indulin, Westvaco Co.), and LS obtained as a by- product of the sulfite pulping process (Nippon Seishi Chemicals). Air-dried samples in powder form (about 7 mg) were heated at 20°C min^{-1} from room temperature to 600°C in a flowing nitrogen atmosphere (200 mL min^{-1}). TG curves indicated that a slight mass decrease occurred at a temperature between 0 and 150°C. This mass decrease is attributed to vaporization of residual water in the sample. The air-dried samples contained a certain amount of bound water whose vaporization was detected at a temperature higher than 100°C [60]. Vaporization was not completed while heating to 100°C, due to the interaction between water molecules and the hydroxyl group of the samples [58, 105]. Thermal decomposition of the samples started at around 200°C. The extrapolated decomposition temperature of KL, HL, and LS at around 321, 324°C respectively. The peak temperature appeared in derivative TG curves of lignin at around 390°C. The residual mass of lignin at 500°C was ca. 45%. This characteristic feature of lignin thermal decomposition in a flowing nitrogen atmosphere indicates that lignin phenolic groups condense during thermal degradation. This condensation during thermal decomposition is considered to be very similar to the process of pyrolysis under nitrogen.

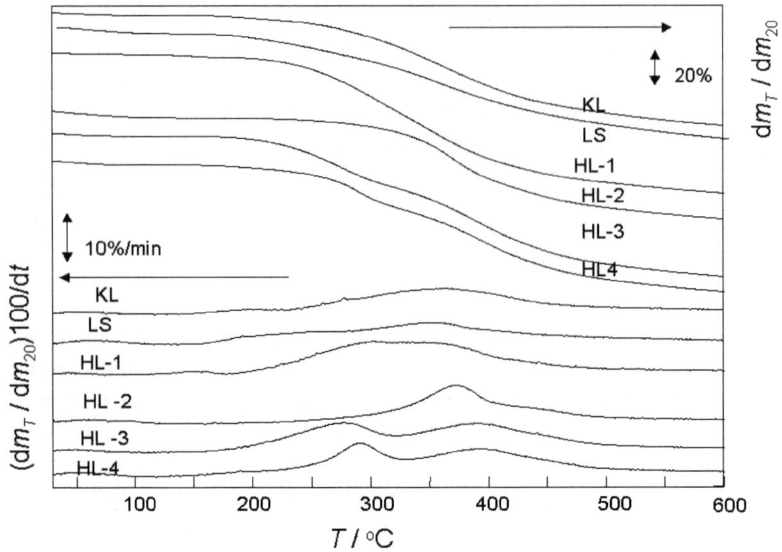

Fig. 14 TG curves and their derivative TG curves of KL, LS and various HLs

5 New Polymers Derived from Chemicals Obtainable by Lignin Decomposition

All currently important industrial organic chemicals such as hydrocarbons, alcohols, polyols, ketones, acids, and phenol derivatives can be obtained by chemical processing of wood [106]. Plastics and synthetic fibres produced could be synthesized using the above chemicals from plant components. In order to develop both current and new polymers, combining petrochemistry with wood chemistry has been considered. In this section, novel polymers synthesized using lignin degradation products are described [107, 108].

5.1 Polyhydroxystyrene Derivatives

Polystyrene derivatives having 4-hydroxyphenyl, guaiacyl, and syringyl groups as shown in Scheme 5 were synthesized [109]. Polystyrene derivatives are amorphous and glass transition was observed by DSC [110]. Due to the presence of the hydroxyl group, glass transition temperature is observed at a higher temperature than that of polystyrene. T_g of poly(4-hydroxy styrene) (42) $\left(M_n = 2.2 \times 10^5, M_w/M_n = 3.2\right)$ was 182°C, poly(3-methoxy, 4-hydroxystyrene) (43) $\left(M_n = 1.1 \times 10^5 M_w/M_n = 4.1\right)$ was 142°C and poly(3, 5-methoxy, 4-hydroxystyrene) (44) $\left(M_n = 9.9 \times 10^4 M_w/M_n = 3.7\right)$ was 108°C [111]. It is clear that intermolecular hydrogen bonding restricts molecular motion and the methoxyl

Scheme 5 Chemical structure of polyhydroxystyrene derivatives

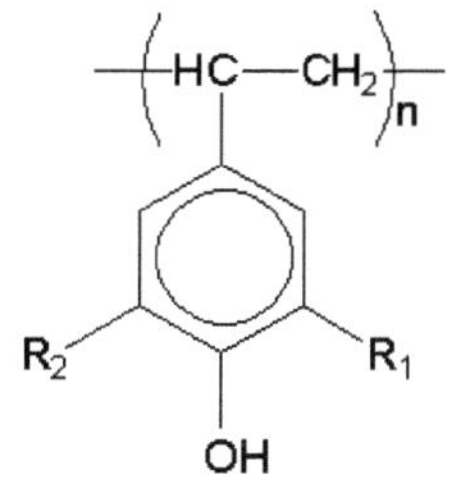

(42)R_1=R_2=H

(43)R_1=OCH_3, R_2=H

(44)R_1=R_2=OCH_3

group enhances the molecular mobility [110, 111]. Hydrogen bonds formed in polyhydroxystyrene derivatives are broken by water molecules and T_g decreases ca 100°C when each hydroxyl group is restrained by one water molecule [55, 112]. Biodegradability of poly(3-methoxy, 4-hydroxystyrene) (43) using extracted bacteria from soil was confirmed and decomposition process was estimated [109]. The hydroxyl and methoxyl group play an important role for biodegradation of polyhydroxystyrene derivatives [109].

5.2 Engineering Plastics

5.2.1 Polyethers

Polyethers (Scheme 6) having 4-hydroxyphenyl or guaiacyl groups, poly(oxy-1,4 phenylenecarbonylmethylene (**45**) and poly(oxy-2, methoxy-1,4-phenylene -carbonylmethylene) (R = OCH_3) (**46**) were synthesized [113]. X-ray diffractogram indicates that poly(oxy-1,4phenylenecarbonylmethylene) is crystalline and poly(oxy-2,methoxy-1, 4-phenylenecarbonylmethylene) is an amorphous polymer. Thermal decomposition of both polymers starts at around 230°C [114]. Aromatic polyethers having phosphine oxide in the main chain were derived from phenols having core structure of lignin. Decomposition temperatures of the above polymers were higher than 500°C. It was found that enthalpy relaxation is markedly affected by introduced functional groups [115, 116].

5.2.2 Polyacylhydroazone Derivatives

Polyacylhydrazone (Scheme 7) derivatives having guaiacyl groups in the repeating unit were synthesized using vanillin and dibromoalkane as starting materials. When

Scheme 6 Scheme of polyethers having 4-hydroxyphenyl or guaiacyl groups

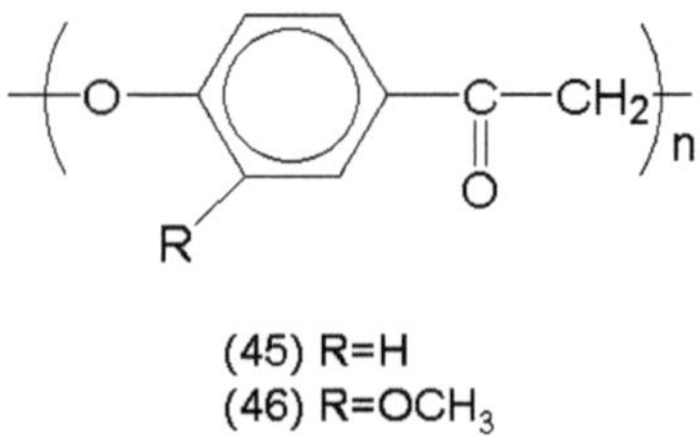

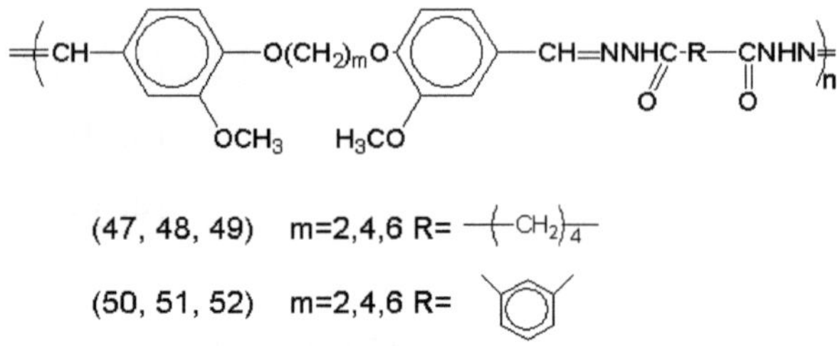

Scheme 7 Polyacylhydroazone having guaiacyl groups

Table 2 T_g and T_d of the polyacylhydrazone derivatives

No	$(CH_2)_m$	R	T_g	T_d
47	2	$(CH_2)_4$	142	336
48	4	$(CH_2)_4$	115	336
49	6	$(CH_2)_4$	107	337
50	2	C_6H_4	211	349
51	4	C_6H_4	194	345
52	6	C_6H_4	193	342

the unit R is the same, molecular flexibility increases with increasing length of alkylene chain. When the number of m is the same, the aromatic groups increase the molecular rigidity [116]. T_g and T_d of the polyacylhydrazone derivatives shown in Scheme 7 are shown in Table 2.

The above results indicate that the molecular mobility of lignin is affected by rigid aromatic ring in the main chain and that molecular flexibility is controlled by paraffinic chains.

5.2.3 Polyesters

Polyesters (Scheme 8) having guaiacyl groups and spiro-dioxane rings in the main chain were prepared using vanillin and pentaerythritol as starting materials. Neither glass transition nor melting was observed for polyester having terephthaloyl group in the main chain, T_d was found at 285°C. In contrast, when the sebacoyl group is introduced T_g was observed at 85°C and T_d at 254°C [117].

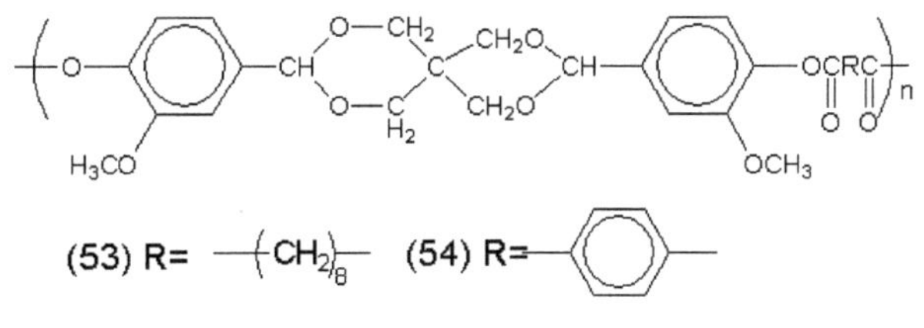

Scheme 8 Polyesters having guaiacyl groups

Scheme 9 Polybenzalazine having guaiacyl groups

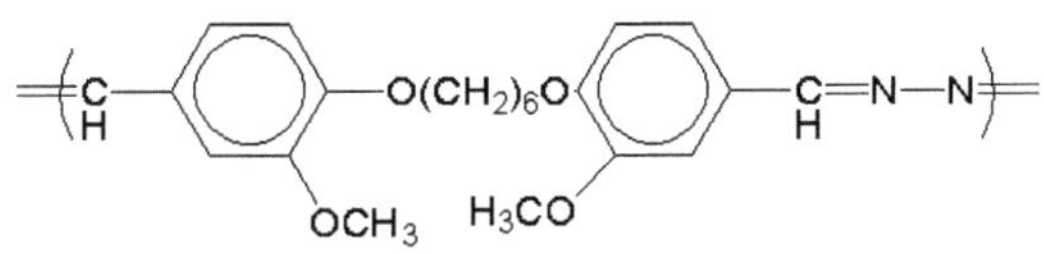

(55)

5.2.4 Polybenzalazine

Polybenzalazine having guaiacyl groups in each repeating unit was prepared (Scheme 9). The obtained polymer was not soluble in dioxane, DMSO, DMF, *m*-cresol or chloroform. Melting temperature was observed at 205°C in a DSC heating curve $\left(10°\mathrm{C\,min}^{-1}\right)$. T_d was ca. 300°C. Polybenzalazine is a stable crystalline polymer [118, 119].

The engineering plastics shown in the above sections and others [120, 121] indicate that chemicals obtained by lignin decomposition (Chap. 4) can be used as starting materials of polymerization. This presents a possibility for the future development of lignin plastic industries, i.e., various amorphous and crystalline polymers can be obtained by introduction of functional groups in lignin structures. Polymers having lignin mimic structure show high T_g and T_d. At the same time, the molecular motion of the above polymers explains the molecular motion of lignin in the solid state. However, at present, practical industrial production is not readily attained since they are not competitive with similar kinds of engineering plastics derived from petroleum because of merchantization. It should also be considered that (1) chemicals from lignin are not produced in industrial levels and (2) mechanical performance has not been thoroughly studied in a large scale.

6 New Polymeric Materials Derived from Industrial Lignins and Related Biomass

Lignin can be obtained in a large scale as a by-product of the pulp industry. A major part of lignin is burnt as a fuel for recovery boilers. A small part of isolated lignin has been used as a filler in concrete, inks, and paints [122, 123], or as a soil stabilizer. Among various types of industrial lignin, LS is water-soluble and used as a dispersant because of amphiphilic properties. When lignin was blended with natural and synthetic biodegradable polymers [124, 125], homogenous blends were scarcely obtained [126]. Glasser and his coworkers carried out chemical modification of lignin [44, 127–130] and various types of copolymers and blends were prepared using modified lignins such as hydroxypropyl lignin [131–143]. It has also been reported that oxypropylation is useful to sythesize polyesters and PU [144]. Recently Funaoka and his coworkers prepared lignin derivatives, such as lignophenol and lignoresol [145]. They attempted to utilize them for various purposes such as dispersants [146], gels as a selective recovery of metal [147–149], a photosensitizer for dye-sensitized solar cell [150].

In order to utilize lignin by converting biocompatible new polymers, it is crucial to use lignin as obtained from the pulp industries. For practical utilization of lignin in a large scale, polymers that have been derived from lignin should have good physico-chemical characteristics, as well as good cost performance, since polymers derived from lignin have to be competitive with other synthetic polymeric materials derived from petroleum. From the above points, the chemical modification process that increases the cost for large scale production of new polymers is not preferable from the viewpoint of practical utilization. Recent advances in lignin utilization for practical domains have focused on the cost performance of lignin resources and the applicability of the present production facilities in order to develop new polymers.

6.1 Polyurethanes

Polyurethane→PU is the most widely used three-dimensional polymer. Various types of PUs produced on a large scale are found in cars, housing, clothing, and medical applications [151, 152]. On this account, if PU derived from petroleum is converted into PU from biocompatible resources, a large mass of lignin could be utilized as a raw material of polymer industry, having an impact on a wide range of industrial fields. In this section, PU preparation from lignin and its physical properties are described.

For the chemical reaction of lignin in the preparation of PU, it is essential that lignin is homogeneously soluble in organic solvents. It has been thought that lignin is soluble in a limited number of solvents and that this causes the difficulty of chemical reaction of lignin. Based on the above understanding, lignin derivatives were first prepared, and then second step chemical reaction was carried out. Pre-chemical procedure [153] on a large scale before PU formation requires new facilities and necessarily reduces the competitive advantage in cost performance with materials derived from petroleum. Accordingly, it is preferable to use lignin without further purification or pre-treatments. In order to attain the PU formation using lignin as obtained from the pulping industry, it is crucial to find a suitable solvent which can be used under mild conditions.

The most important finding was that various kinds of industrial lignin are soluble in a series of polyethylene glycol with a wide range of molecular mass from

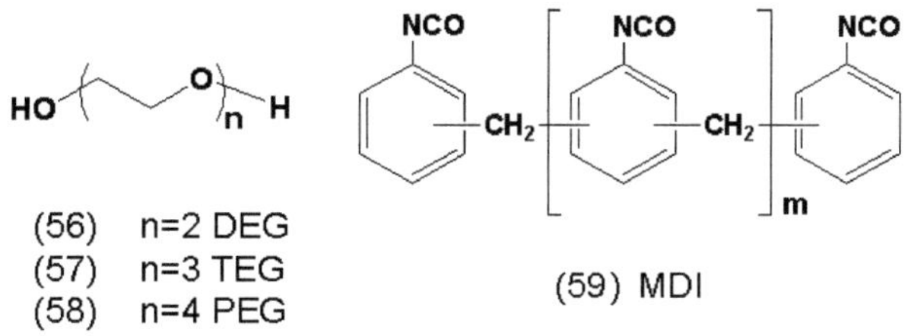

Scheme 10 Chemical structures of DEG, TEG, PEG200 and MDI

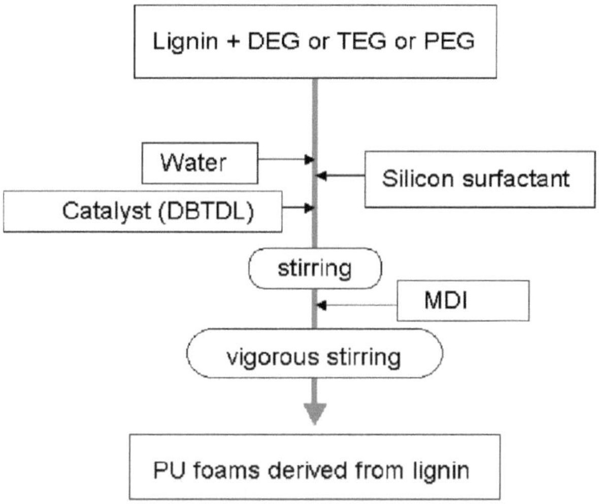

Fig. 15 Preparation of lignin-based rigid PU foams *HL-1* New Zealand; *HL-2* Russia (Archangelsk plant); *HL-3* Russia (Lobinsky plant); *HL-4* Russia (Kirovsky plant)

ca.70 to 3,000 [42, 154–156]. As raw materials for the preparation of lignin-based PUs, various kinds of industrial lignins, such as KL from kraft pulping waste liquor, LS from sulfite pulping waste liquor, and HL which was obtained as a by-product of bio-ethanol from wood meal, can be used without further pre-treatment. Those lignins were dissolved in polyols such as DEG (Scheme 10, (**56**)), triethylene glycol→TEG (**57**), and polypropylene glycol (PPG) (**58**) under heating from 60 to 70°C in order to prepare polyol solutions containing lignin. The above polyol solutions with various lignin contents of 0–33% were mixed with poly(phenylene methylene) polyisocyanate (MDI) (**59**) at room temperatures, and polyurethanes were prepared. In order to prepare PU foams, one of the above lignin-based polyol solutions was mixed with a plasticizer, surfactant (silicon oil), and a catalyst (di-*n*-butyltin dilaurate, DBTDL), and then MDI was added. This mixture was stirred with a trace amount of water which was added as a foaming agent. NCO/OH ratio was 1.0–1.2 [156]. Preparation of lignin-based rigid polyurethane foams is shown in Fig. 15. Chemical structures of DEG, TEG, polyethylene gycol (PEG 200), and MDI are shown in Scheme 10. Scheme 11 shows the schematic chemical structure of lignin-based PU.

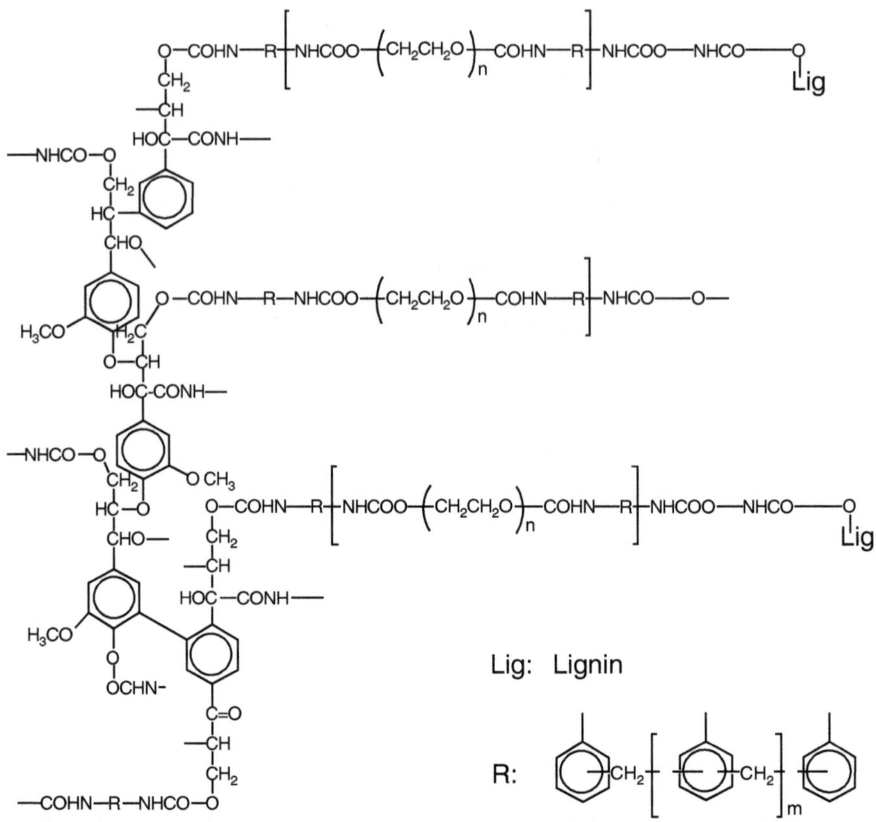

(60)

Scheme 11 Schematic chemical structure of lignin-based PU

6.1.1 Rigid Polyurethane Foams Derived from Kraft Lignin

PU foams derived from KL (KLPU foams) were prepared according to the methods mentioned above [42]. PU foams obtained from the above three kinds of lignin-based DEG, TEG, and PEG solutions are designated as KLDPU, KLTPU, and KLPPU, respectively. Apparent density→ r(ρ) is an index of characteristic feature of PU foams. Generally ρ values decrease with increasing KL content regardless of kinds of polyol. This suggests that the cell size increases with increasing KL content due to the increase of viscosity of KL polyol.

T_g of the above PU samples was measured by DSC. Figure 16 shows a three-dimensional figure between T_g,, KL content and number of repeating unit of oxyethylene chain (n) of KLDPU, KLTPU and KLPPU. As shown in Fig. 16 the molecular chain length of oxyethylene group markedly affects molecular mobility of PU. In contrast, the amount of KL content scarcely affects the T_g, values in the above PU derivatives.

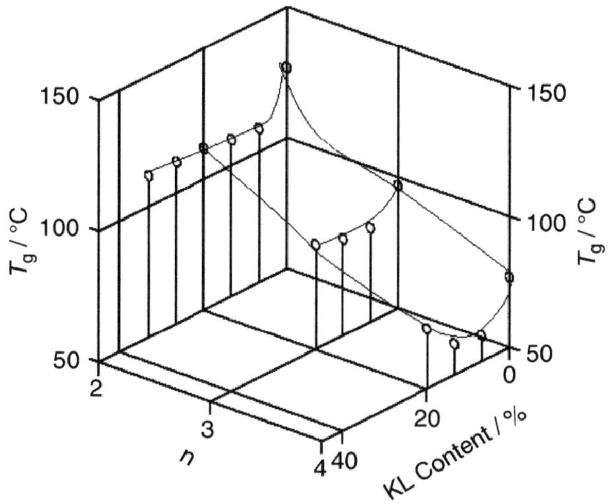

Fig. 16 Three-dimensional figure between T_g, KL content (%) and number of repeating unit of oxyethylene chain (*n*)

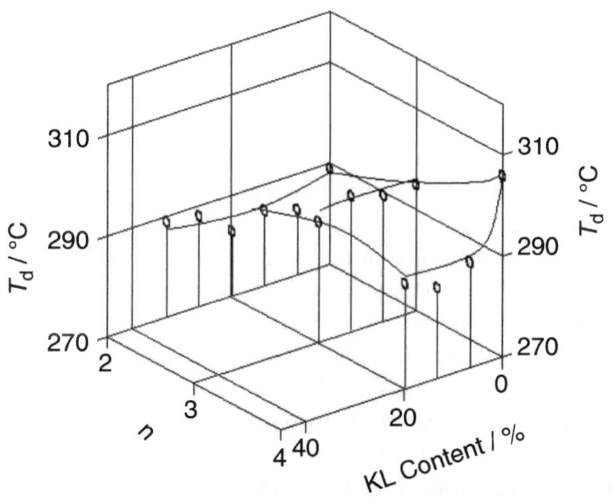

Fig. 17 Relationships between T_{d1}, *n* and KL content of PU foams derived from KL

Decomposition temperature→T_d of the above KL foams was investigated by TG. Thermal decomposition was observed as two stages in TG curves for all PU foams. The low temperature side T_d (T_{d1}) observed at around 290°C is the major decomposition and is attributed to thermal degradation of urethane bonding formed by reaction of phenolic hydroxyl and isocyanate groups. The high temperature side T_d (T_{d2}) is attributed to that of urethane bonding between alcoholic hydroxyl groups and isocyanate groups [101]. T_{d1} slightly increases with increasing n as shown in Fig. 17. In KLDPU and KLTPU, T_{d1} does not change obviously with increasing KL content. However, T_{d1} of KLPPU decreases slightly with increasing KL content.

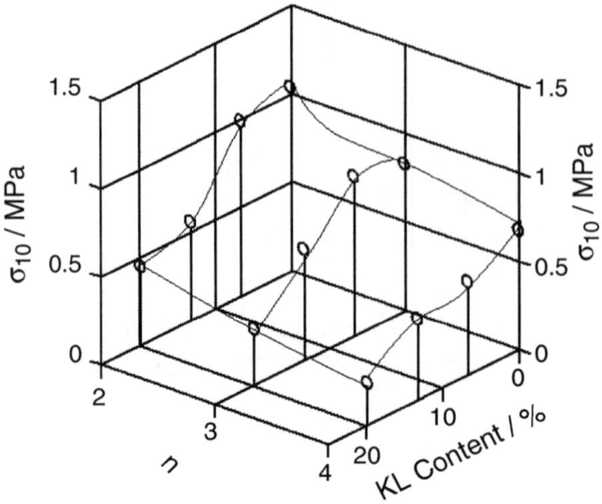

Fig. 18 Relationships between σ_{10}, *n* and KL content of PU foams derived from KL

Since the increase of KL content in KLPPU reduces the amount of PEG 200 (PEG having M_w = ca. 200), the dissociation between the phenolic hydroxyl group in KL and the isocyanate group becomes prominent.

Mechanical properties of KLPU foams were measured by compression tests. The compression strength at 10% strain is designated as σ_{10} (MPa). The yielding point was also observed at around 4–5% strain for all rigid PU foams. Compression modulus (E) was calculated from the initial slope of compression curves. As shown in Fig. 18, the values of σ_{10} decrease with increasing KL content and length of oxyethylene chains. This indicates that the compression strength and compression elasticity of rigid polyurethane foams highly dependent on the values of ρ.

Compression modulus (E) was calculated from the initial slope of compression curves. Figure 19 shows relationships between E, n and KL content of KLPU foams. The E values vary in a similar manner with σ_{10}.

Figure 20 shows the relationship between σ_{10} and ρ As shown in Fig. 20, a linear relationship is obtained regardless of the number of oxyethylene chains, although data scatter in a certain range.

6.1.2 Rigid Polyurethane Foams Derived from Sodium Lignosulfonate

LS is obtained as a by-product of sulfite pulping process. LS has not been used as a chemical component of common lignin-based polymers such as epoxy resins due to its ionic nature. LS is soluble in water and, at the same time, it is soluble in some kinds of solvent such as ethylene glycol, DEG, TEG, and PEG. [157]. LS-based PU foams can be prepared from LS-based polyols, LSD derived from DEG, LST from TEG, and LSP from PEG. The thermal and mechanical properties of rigid PU foams

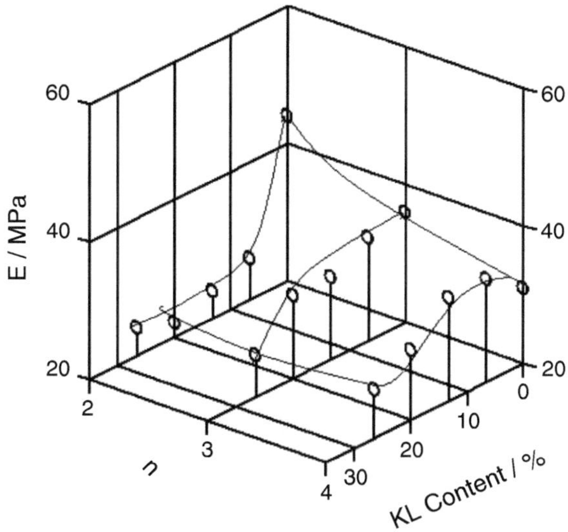

Fig. 19 Relationships between E, n and KL content of PU foams derived from KL

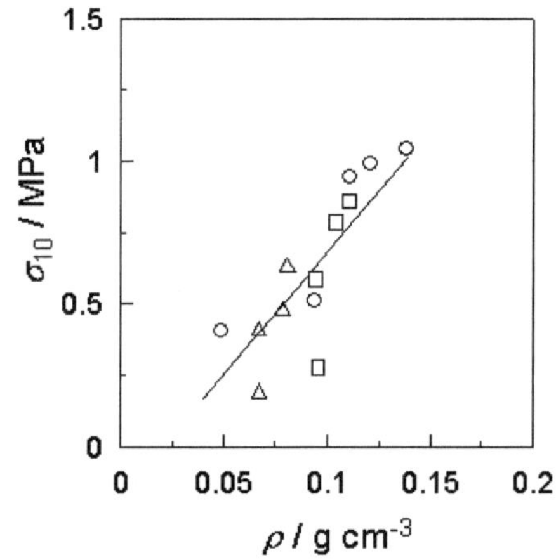

Fig. 20 Relationship between compression strength at 10% strain (σ_{10}) and ρ; *open circle*, KLDPU; *open square*, KLTPU; *open triangle*, KLPPU

derived from LSD, LST, and LSP that were investigated by DSC, thermogravimetry TG, TG-Fourier transform infrared spectroscopy (TG-FTIR) and compression tests, have been described [101].

Figure 21 shows relationships between T_g, n and LS content of LSPU foams. It is clearly seen that the T_g value is affected by the chain length of oxyethylene chain. The amount of LS scarcely affects T_g values, especially when n value is small. In the case of LSPPU, T_g increases from 80 to 90°C with increasing LS content, since the rigid phenylpropane structure in LS acts as a hard segment efficiently in PU networks containing long oxyethylene chains of PEG.

Thermal degradation of LSDPU was observed in two stages. The low temperature side thermal degradation (T_{d1}) was observed at around 290–300°C depending

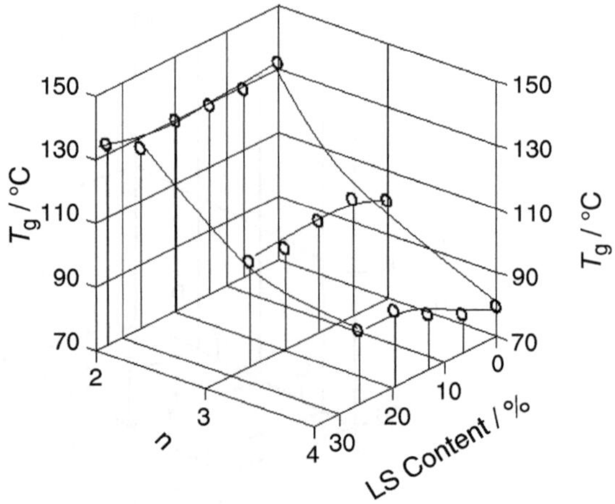

Fig. 21 Relationships between T_g, n and LS content of LSPU foams

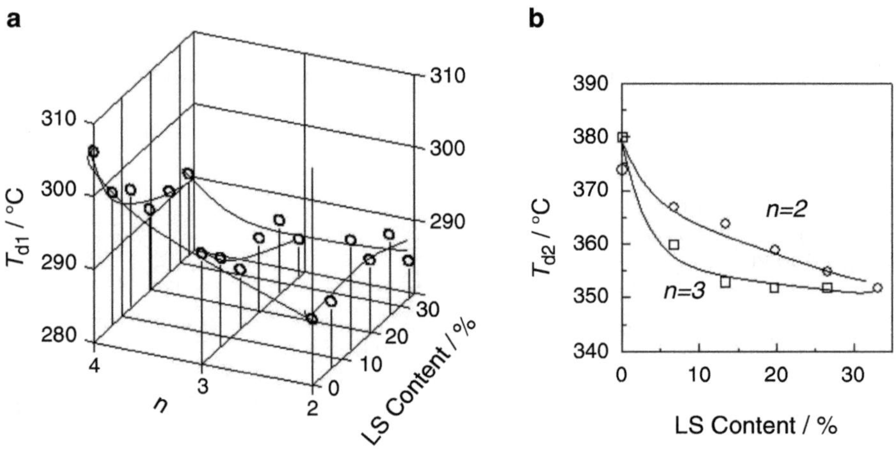

Fig. 22 Relationships between change of T_{d1}, n and LS contents in polyol (**a**) and relationship between T_{d2}.and LS content (**b**)

on n and LS content, as shown in Fig. 22a. The mass change of the low temperature side degradation was far larger than that of high temperatue side degradation. T_{d2} decreased noticeably with increasing LS content. In this series of LSPPU, the high temperature side degradation tempeature (T_{d2}) was not distinctly covered by the tail of the low temperature side mass change, as shown in Fig. 22b. It is known that the dissociation of urethane bonding between the phenolic hydroxyl group and the isocyanate group occurs at around 250°C, which is lower than that of urethane bonding between the alcoholic hydroxyl group and the isocyanate group [42, 101]. The increase of LS content in the above LSPUs reduces the amount of polyols such as DEG, TEG, and PEG, which are thermally stable compared with LS, when they form urethane bonding with the reaction of isocyanate.

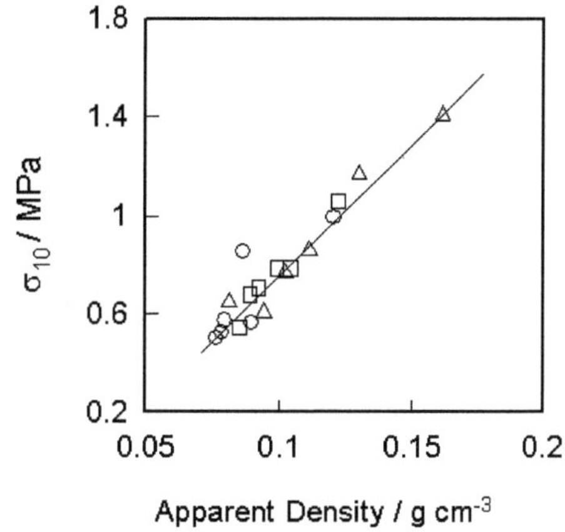

Fig. 23 Relationship between compression strength at 10% strain (σ_{10}) and apparent density (ρ) of LSPU foams. *Open circle*, LSDPU; *open square*, LSTPU; *open triangle*, LSPPU

Compression strength at 10% strain (σ_{10}) and compression modulus (E) of LSPU foams were obtained by mechanical test. The values of σ_{10} and E are markedly affected by the apparent density (ρ) of PUs. The apparent density could be controlled by foaming conditions, such as viscosity of the reaction mixture and the amount of foaming reagent. As shown in Fig. 23, the value of σ_{10} linearly increases with increasing ρ. A similar relationship was observed in the case of compression modulus (E). The above results clearly indicate that mechanical properties of rigid PU foams are dependent on ρ.

6.1.3 Rigid Polyurethane Foams Derived from Hydrolysis Lignins

Four kinds of hydrolysis lignin (HL) samples obtained from different industrial plants in New Zealand and three different places in Russia (Archangelsk, Lobinsky, and Kirovsky) were used. Lignin powder samples were dissolved in DEG or polyethylene glycol (PEG) and HL polyols were prepared [54]. Rigid polyurethane (PU) foams were obtained through the reaction of HL polyols with MDI. Optimum reaction conditions in order to attain designed properties were examined.

As shown in Fig. 24, a thin cell wall was homogeneously observed. All of PU foams derived from both HL showed similar pore size as shown in Fig. 24. It is clear that PU foams contain a large vacancy. Image analysis of micrographs was carried out and pore size distribution was calculated.

The apparent density (ρ) of PU foams derived from HL polyols were in a range of 0.04–0.075 g cm^{-3}. Compression strength of the above PU foams was in a range 0.2–1.0 MPa and modulus of elasticity in the range 5–10 MPa. Glass transition temperature can be varied from 80 to 120°C. Thermal conductivity of the foam was 0.032–0.037 W m^{-1} K^{-1} which was smaller than in the usual commercial PU foams derived from petroleum. All the above physical properties and the cost-performance are better or compatible with commercial PU products.

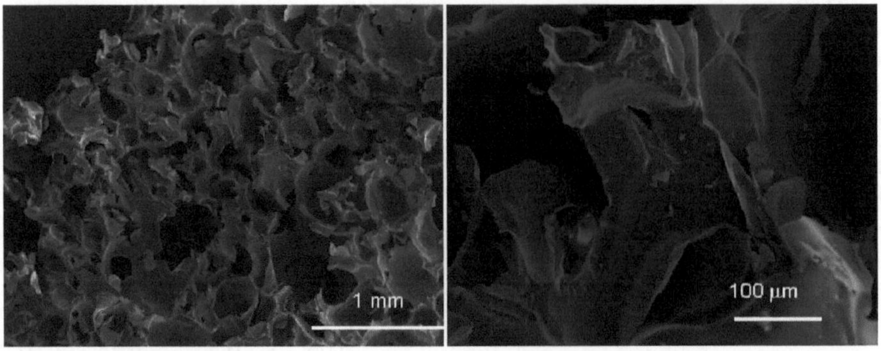

Fig. 24 SEM of PU foams derived from HL-2 (Russia, Lobinsky plant)

6.1.4 Polyurethane Foams Derived from Lignin- and Molasses Mixtures

Molasses→ML (ML) is obtained as a by-product of the sugar industry. The major components of ML are saccharides mainly mixed with sucrose, glucose, and fructose [158]. As already mentioned in the previous sections, LS obtained from the sulfite pulping process is a polyelectrolyte, which is soluble in water and a certain number of organic solvents. Both lignin and molasses are obtained as industrial residues of large-scale industries and they have not been utilized in a broad range of practical fields, although they are the most promising natural resources. In order to control the physical properties of PU, ML-based polyols were prepared and mixed with lignin polyols in a broad range of mixing ratios [159–161].

The photographs in Fig. 25a show PU foams derived from LSP/ML polyol derived from PEG (MLP) mixtures that are yellowish brown, and color becomes deeper with increasing LS content. Apparent density is in the range 0.03–0.07 $\mathrm{g\,cm^{-3}}$. Apparent density (ρ) increases with increasing NCO/OH ratio which is an index of cross-linking and slightly increases with LSD content (Fig. 25b). The values of ρ are almost similar to those of PU foams derived from HL polyol derived from DEG.

By compression test of PU foams derived from LSD/MLD polyol mixture, the compression strength at 10% strain (σ_{10}) and compression modulus were calculated from stress-strain curves. As shown in Fig. 26, σ_{10} values increase with increasing NCO/OH ratio. At the same time, an optimum condition seems to exist at around LSD/MLD = 70/30. This suggests that MLD takes crucial role in order to control the mechanical properties.

Figure 27 shows that mechanical properties of PU foams linearly correlate with apparent density. This indicates that mechanical properties can be predicted from apparent density value. In order to choose reaction conditions to design the PU foams having suitable properties, apparent density is a suitable index.

As mentioned in the previous sections, it is clear that the molecular mobility increases with increasing the number of oxyethylene unit→ n (n). [157]. Thus, T_g decreases with increasing molecular mass of ethylene glycol. For the development

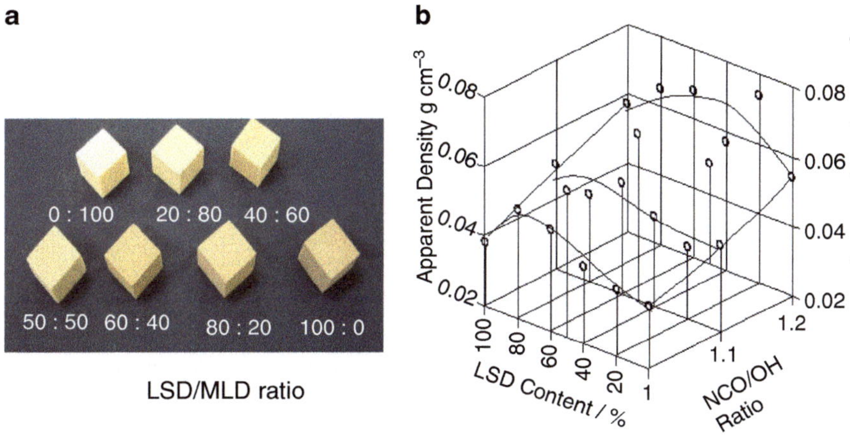

Fig. 25 Photographs of PU foams derived from LSD/MLD mixed polyol (**a**) and relationships between apparent density, LSD content and NCO/OH ratio (**b**)

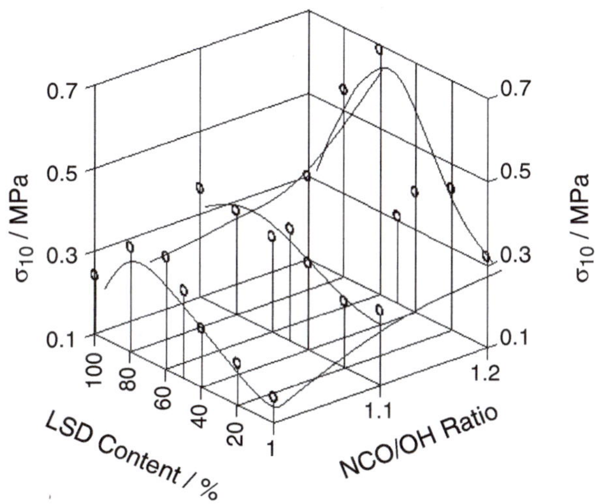

Fig. 26 Relationships between σ_{10}, LSD content and NCO/OH ratio of PU foams derived from LSD/MLD mixtures

of insulators for housing or other purposes, T_g of PU should be higher than the atmospheric temperature. In order to obtain rigid PUs DEG was chosen as a solvent of LS and ML. Figure 28 shows change of T_g with NCO/OH ratio and LSD/MLD ratio. T_g was observed in a temperature range 75–120°C. Mixing of polyols is also important to stabilize the cross-linking reaction, although the effect of LSD/MLD ratio on T_g was not distinct. From DSC data, it can be said that the rigid phenylpropane ring in lignin and the glucopyranose in saccharides ring similarly contribute to the main chain motion of PU, while cross-linking density markedly affects the T_g.

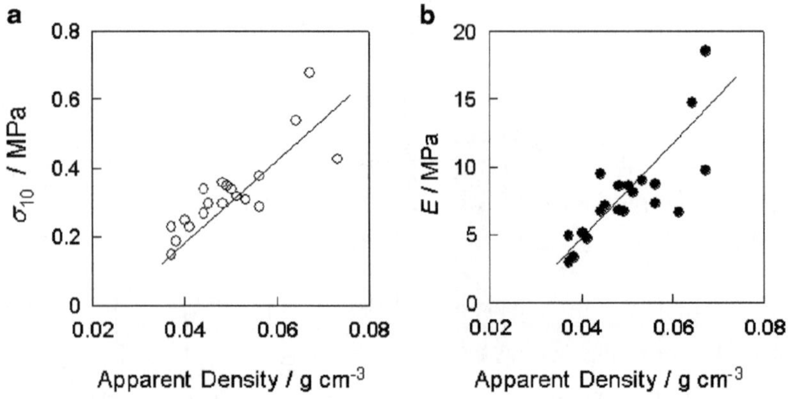

Fig. 27 (**a**) Relationship between σ_{10}, compression modulus (E) and (**b**) Apparent density of PU foams derived from LSD/MLD mixtures

Fig. 28 Change of glass transition temperature (T_g) with NCO/OH ratio and LSD content

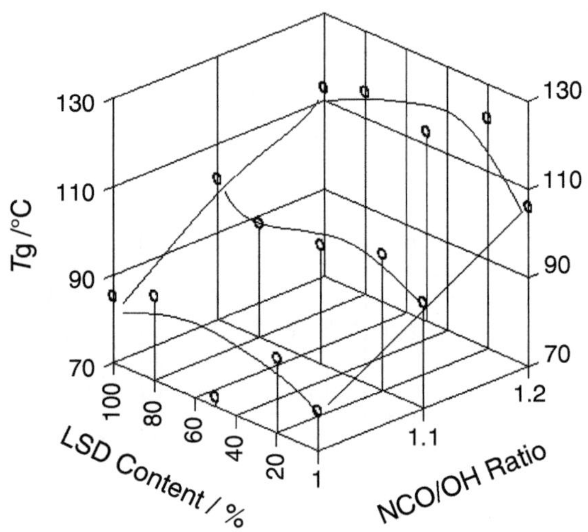

TG and DTG curves of PU foams derived from molasses and lignin with various MLD/LSD ratios were measured in N_2 atmosphere. The mass of PU foams decreases in the two steps. The low temperature side decomposition (T_{d1}) is in a temperature from 285 to 295°C depending slightly on NCO/OH ratio and MLD/LSD ratio. The high temperature side decomposition (T_{d2}) is observed at temperatures from 355 to 365°C. From TG and DTG curves, it was observed that the major part of PU foams decomposes in the first step. The mass residue at 500°C was in the range 40–45%, regardless of NCO/OH ratio or MLD/LSD ratio. Compared with synthetic polymers, biopolymers such as lignin and polysaccharides are thermally stable and temperature range of thermal decomposition is wide. The amount of residual component, mainly charcoal, is 20–40% depending on plant species, portion of plant, extraction method. When other chemical components are introduced

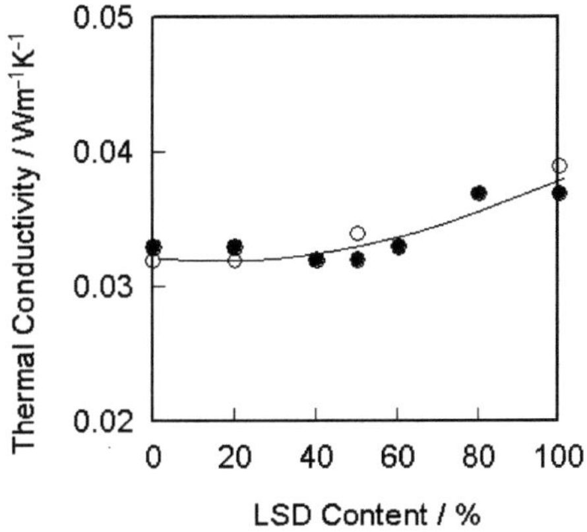

Fig. 29 Relationship between thermal conductivity and LSD content. NCO/OH ratio *closed circle* = 1.2, *open circle* = 1.1

Table 3 Comparison of thermal conductivity of PU foams prepared in this study and commercially obtained foams

Sample	Thermal conductivity $(\mathbf{js^{-1}m^{-1}K^{-1}})$
Polyurethane foam (commercial 1)	0.136
Polystyrene foam (commercial 2)	0.035
Lignin/Molasses foam	0.037–0.032

to biopolymers by chemical reaction, thermal stability generally decreases. This strongly indicates that the higher order structure of biopolymers that evolved in the long term attained adaptable characteristics in nature.

In order to utilize PU foams for insulation materials, it is necessary to know the thermal conductivity of the sample. Thermal conductivity was measured by sandwiching the PU foam between two plates, whose temperature difference was pre-determined (in this case 27.7°C). Figure 29 shows the relationship between thermal conductivity and LSD content. Time maintained for the equilibration did not affect the results, when time exceeded 20 min. With increasing LSD content, thermal conductivity increased slightly. The above results were compared with commercial grade polyurethane and polystyrene foams. The results are shown in Table 3. The data suggests that the thermal conductivity of PU foams derived from lignin and molasses can be used as alternative materials derived from petroleum. The results indicate that PU foams derived from biopolymers can be used as insulation materials.

6.2 *Polycaprolactone Derivatives Derived from Lignin*

Polycaprolactone (PCL) is a representative biodegradable polymer. The biodegradability of PCL has been investigated (for example, [162]). PCL is a crystalline polymer whose melting is observed at around 60°C, although thermal degradation temperature is sufficiently high among synthetic polymers. This indicates that the temperature range where PCL in the solid state is too narrow for practical utilization. Lignin is a biodegradable polymer whose degradation mechanism has been thoroughly investigated. The rate of biodegradation of lignin in nature is not fast compared with that of polysaccharides or PCL. In contrast, amorphous lignin is thermally stable in a wide temperature range. Accordingly, it is considered that lignin combined with PCL is an interesting research subject [163, 164].

Lignin-based polycaprolactone graft copolymers were prepared from two different types of lignin, KL and LS. In order to vary the length of grafted PCL chains, CL/OH ratio was changed from 1 to 100 mol mol^{-1}. LS-based PCL (LSPCL) and KL-based PCL (KLPCL) were synthesized by polymerization of caprolactone (CL) which was initiated by the hydroxyl (OH) group in the above lignins. The polymerization was carried out for 12 hr at 150–160°C with the presence of a small amount of DBTDL. LSPCL and KLPCL sheets were prepared by heat-pressing the synthesized polymers at 120°C under ca. 10 MPa. A schematic chemical structure of LigPCL is shown in Scheme 12.

FTIR spectra of representative LSPCL samples were measured. The absorption band at 1,722 cm^{-1} attributed to the stretching band of $C = O$ of PCL chains increases with increasing CL/OH ratio. In order to make a comparison among different IR spectra, the absorption band at 1,593 cm^{-1} attributed to the phenyl group of lignin was used as an inner standard and the ratio of $A_{1722\,cm-1}/A_{1593\,cm-1}$ for each spectrum was calculated.

Figure 30 shows the stacked DSC heating curves of LSPCL with various CL/OH ratios from 10 to 100 mol mol^{-1}. It is assumed that all side chain molecules reacted in a similar manner, and that molecular mass is in a range 1.1×10^3–1.2×10^4. As shown in Fig. 30, marked endothermic peaks due to melting are observed. When CL/OH ratio is over 10 mol mol^{-1}, marked endothermic peaks due to melting were also observed in KLPCL. The above results suggest that both LS- and KL PCLs with CL/OH ratios over 10 mol mol^{-1} have a clear crystalline region in the molecular structure. Melting temperature is maintained at almost the same temperature,

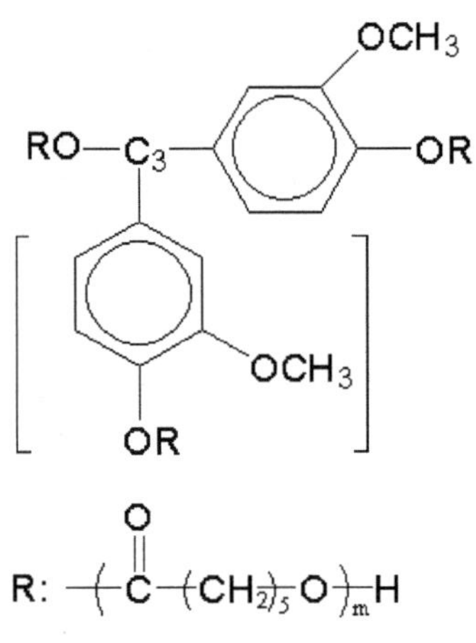

(61)

Scheme 12 Schematic chemical structure of lignin-based PCL

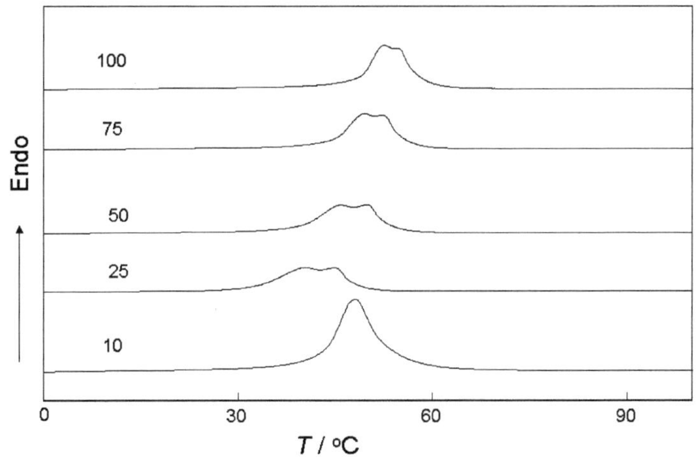

Fig. 30 DSC heating curves showing melting region of LSPCL with various molecular mass

Fig. 31 Relationships between T_g and CL/OH ratio of LSPCL. *Closed circle*, batch 1; *open circle*, batch 2

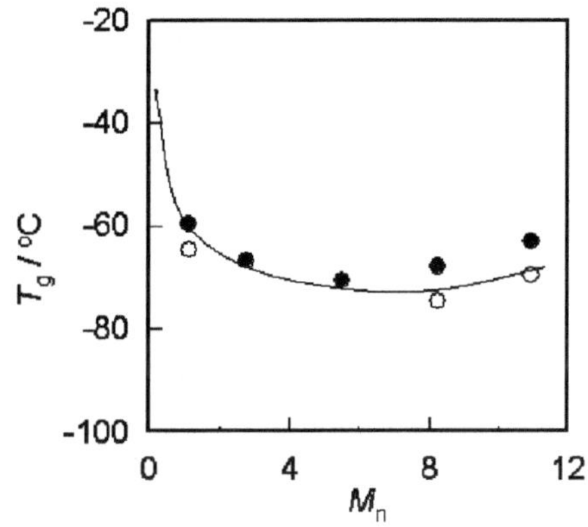

regardless of CL/OH. From melting enthalpy (ΔH_m), crystallinity can be calculated assuming that ΔH_m of 100% crystalline PCL is $139.5\,J\,g^{-1}$. The crystallinity of LSPCL ratio shown in Fig. 30 was 45–49%, regardless of CL/OH ratio when CL/OH ratio exceeded 10. The above results indicate that PCL side chains form a crystalline region without the effect of lignin which acts as an initiator of the ring opening reaction.

When the molecular mass is less than 1×10^3, the crystalline region was not formed readily. Although DSC curves are not shown here, in the low temperature at around -70 to -60°C, the baseline change was observed. Glass transition of lignin is ordinarily observed at around 120–180°C as mentioned in Sect. 3. Thus, T_g observed at around -70 to -60°C, is attributable to the molecular motion of amorphous region of PCL. Beside glass transition, an exothermic peak due to cold-crystallization was recognized in a CL/OH ratio less than 10. Figure 31 shows relationships between T_g and CL/OH ratio of LSPCL (batch 1 and 2).

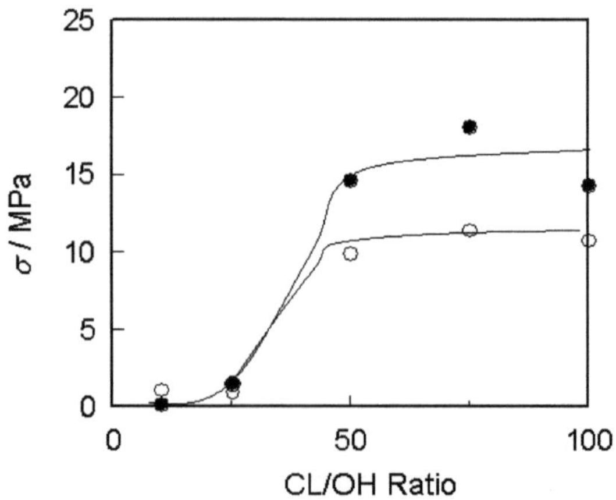

Fig. 32 Relationship between σ and CL/OH ratio of LSPCL and HLPCL. *Closed circle*, HLPCL; *open circle*, LSPCL

Thermal degradation of LSPCL samples was investigated by TG. T_d of LSPCL was observed at around 395°C, regardless of CL.OH ratio. In contrast, T_d of HL-based polycaprolactone (HLPCL) at CL/OH ratio was 370°C and then reached 395°C with increasing CL/OH ratio. Both PCL samples showed one stage thermal degradation which is attributable to decomposition of PCL chains. LSPCL and HLPCL sheets were prepared by heat press under pressure and mechanical measurements were carried out. Figure 32 shows strength of PCL sheets obtained by stretching tests. Values of σ reached a constant value when CL/OH ratio exceeds 50.

From the above results, it is confirmed that LS can be reacted with CL in a similar way to other types of lignin, such as KL and alcoholysis lignin. It is thought that process ability of lignins will be attained by introduction of hydrophobic side chains such as PCL to lignin. Copolymers obtained are expected to have biocompatibility since both lignin and PCL are biocompatible.

6.3 Biocomposites

Biocompatibility of composites has recently attracted considerable interest, since composites are widely used in various fields, such as construction. Waste composites are ordinarily difficult to decompose after use by either mechanical or thermal treatments. The characteristic features of composites, i.e. the behavior and special effects associated with a wide class of heterogeneous media, are the main reasons for this difficulty of decomposition. Components prepared from plastics derived from petroleum are stable in nature. Accordingly, it is necessary to develop new types of composites having better performance that are not only compatible with nature but also competitive in the market.

In this section, PU composites are designed using plant components in both matrix and filler, i.e. two types of lignin-polyol derived from KL and LS are used to prepare polyurethane matrix and wood powder as fillers [165, 166]. In order to increase biodegradability, it is necessary to develop densely filled composites. Mechanical properties of PU composites sandwiched between veneer plates were investigated which are designed to be used as panels having biocompatibility and applications in the field of construction. At the same time, lignin PU composites filled with inorganic fillers was prepared in order to increase thermal stability for practical use as housing materials. Lignin PU applied for use in civil construction is also introduced. Furthermore, lignin PU composites utilizing fillers obtained as residue from agriculture, food and textile industries are described.

6.3.1 Lignin- Molasses-Based PU Biocomposites Filled with Wood Powder

LSD and MLD mixtures were reacted with MDI that was used for the preparation of lignin- and molasses- based PU composites filled with wood powder. For the preparation of the above composites, first wood powder was added to the mixture of LSD and MLD. Then the mixture was stirred at room temperature, MDI was added, poured into a stainless steel frame, and pressed under pressure of ca. 20 MPa at 120°C for 2 hr. In the above biocomposites, wood powder contents were changed from 50 to 90%. LSD and MLD mixture ratios were changed as follows: 0/100, 20/80, 40/60, 50/50, 60/40, 80/20, and 100/0.

Figure 33 shows photographs of PU composites derived from LSD/MLD mixtures (50/50) filled with various amount of wood powder. Dark brown comes from LS component.

Figure 34 shows the relationship between apparent density, wood powder content and LSD/MLS content of PU composites. The maximum values are observed at around wood powder content 70%. This corresponds to the closed packed conditions of fillers. The strength obtained by bending test using three point supporting method shows similar variation as shown in Fig. 35. The maximum values are found from 20 to 25 MPa which depends slightly on LSD content.

The PU composites were examined by thermogravimetry (TG). Sample mass was ca. 7 mg, heating rate was 20°C min^{-1}, N_2 gas flow rate was 200 mL min^{-1} and temperature range 20–600°C. Figure 36 shows TG and TG derivative curves (DTG)

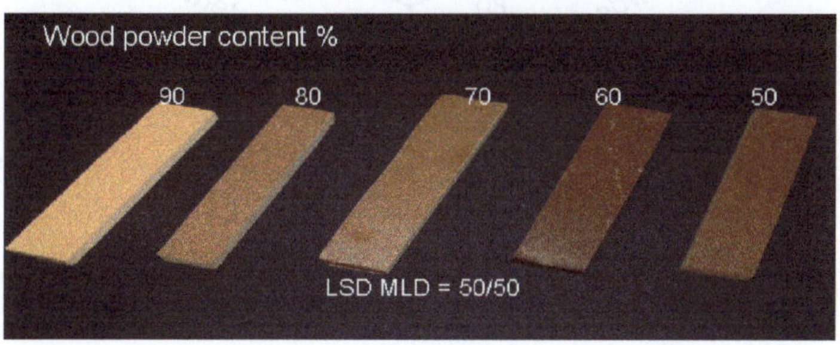

Fig. 33 Photographs of PU composites derived from LSD/MLD mixtures filled with various amounts of wood powder

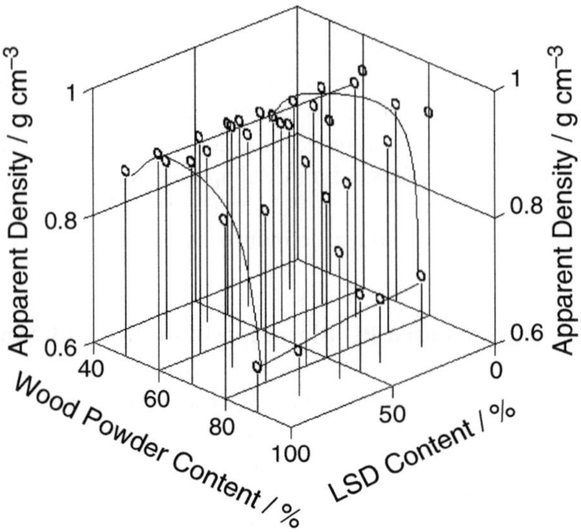

Fig. 34 Relationship between apparent density, wood powder content and LSD/MLS content of PU composites

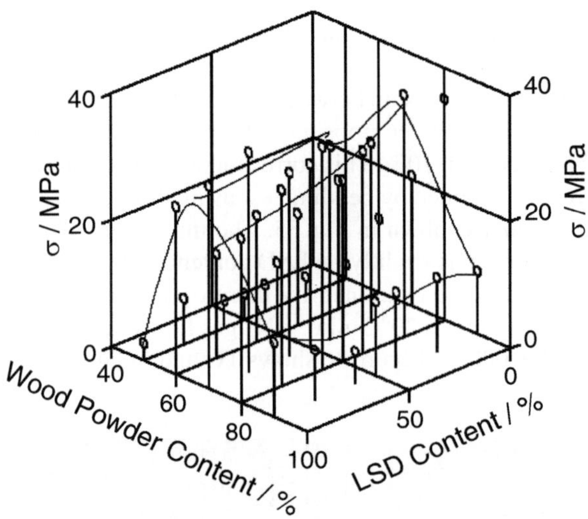

Fig. 35 Relationship between bending strength, wood powder content and LSD/MLS content of PU composites

of PU composites together with various components consisting of PU. TG, DTG curves of cellulose powder are also shown as a reference. Thermal degradations (T_d) of lignin- and molasses-based polyurethane (PU) composites were observed as two steps, low temperature side (T_{d1}) and high temperature side of thermal degradation (T_{d2}). The T_{d1} at ca. 290°C is attributed to the degradation of MLPU and LSPU

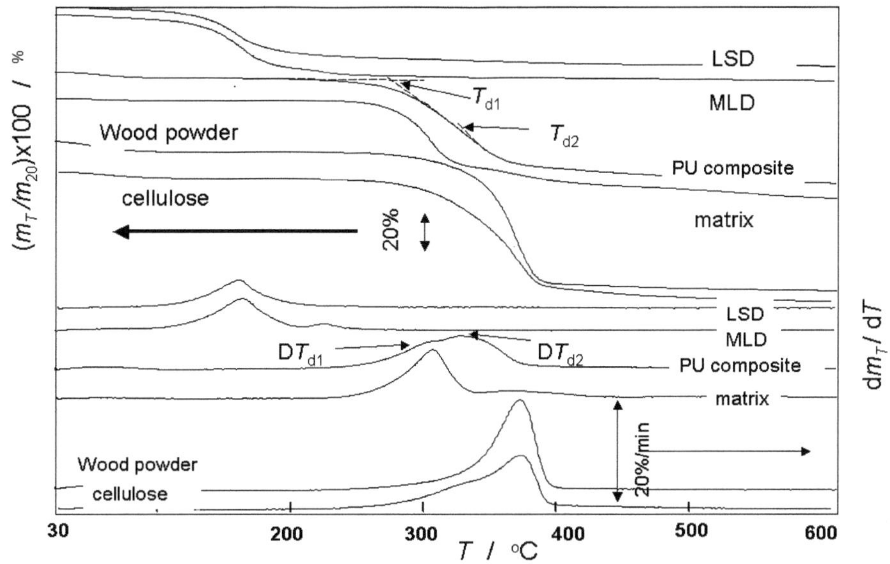

Fig. 36 TG curves of PU matrix derived from LS, ML, wood powder, cellulose powder and PU composites derived from LSD/MLS mixture filled with wood powder

matrix. When the wood powder content increased, T_{d2} observed at 310–360°C shifted to the higher temperature side. The T_{d2} is attributed to the degradation of wood powder. Both degradation temperatures were scarcely affected by the ratio of LSD/MLD. Mass residue (MR) at 450, 500, and 550°C of lignin- and molasses-based polyurethane (PU) composites were examined. MR at 450 and 500°C ranged from 35 to 40% and MR at 550°C from 30 to 35%, regardless of either wood powder content or LSD/MLD ratio.

6.3.2 Wood Panel Derived from Lignin-Based Polyurethane Composites

Wood panels were prepared by sandwiching PU composites filled with wood powder between veneer plates. By using PU composites from lignin, the aim is to develop composites having biocompatibility which can be widely used in various fields, such as construction. Two kinds of lignin KL and LS, and three kinds of polyols having different molecular weight, KLD, KLT, and KLP, were used for the preparation.

Figure 37 shows the side view of a representative wood panel in which PU composite filled with wood powder was sandwiched by veneer plates. The samples filled with wood powder are light brown. In this sample, KLD was used as a polyol. PU composites became slightly darker when LS polyols were used.

Figure 38 shows the change of apparent density of PU composites filled with wood powder as a function of filler (wood powder) content. As clearly seen in

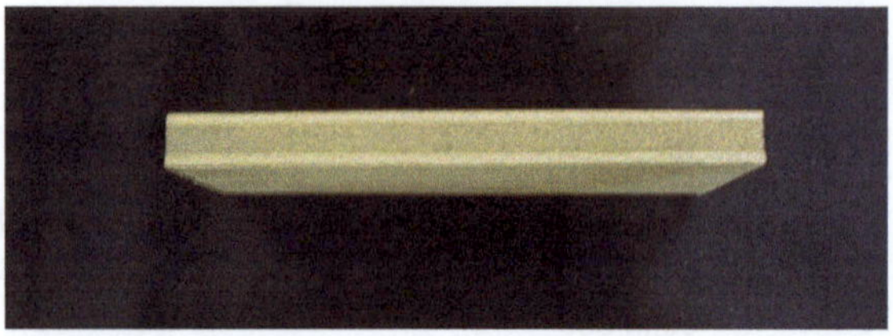

Fig. 37 Side view of representative panel of PU composite filled with wood powder

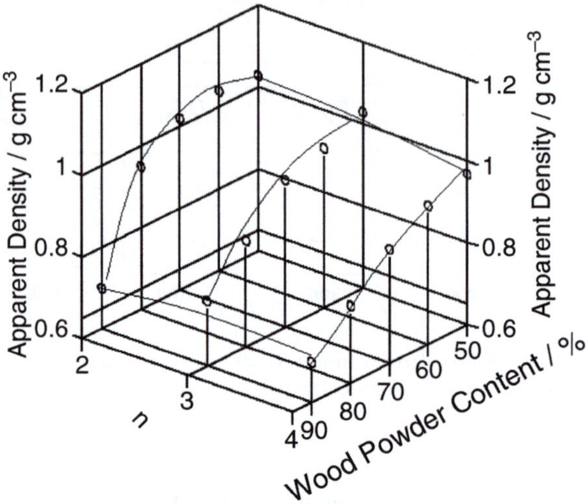

Fig. 38 Relationships between apparent density, number oxyethylene group and wood powder content of PU composites derived from KL

Fig. 38, the apparent densities decrease with increasing wood powder content. The effect of molecular weight of ethylene glycol was more prominent in KL polyols than in LS whose results are not shown here. Apparent density of veneer was $0.54\,\mathrm{g\,cm^{-3}}$ and that of composites $1.2\,\mathrm{g\,cm^{-3}}$; accordingly, apparent densities of the PU composites filled with wood powder sandwiched by two veneer plates were smaller than those of PU composites. Figure 39 shows apparent density obtained by two samples, i.e. PU composite with veneer plates and those that were prepared directly without plates in the same procedure as described in the experimental section. KLD was used for the above two series of samples. Calculated values assuming the summation of each volume of venire plates and sandwiched composite are also shown in Fig. 39. It is clearly seen that calculated values are almost similar to those of samples prepared without plates. These results indicate, foaming reaction took place regardless of veneer plates.

Figure 40 shows the maximum stress divided by apparent density as a function of wood powder content. Values of $\sigma\rho^{-1}$ of PU composite panels derived from KL and LS polyols varied in a similar manner. The maximum was found at around wood powder content 70%, although data was scattered to a certain extent. The effects of

Fig. 39 Relationships between apparent density (ρ) of PU composites with veneers (*closed circle*), calculated PU composites (*open circle*), PU composite prepared without veneers (*open square*) and wood powder content. KLD was used as a polyol

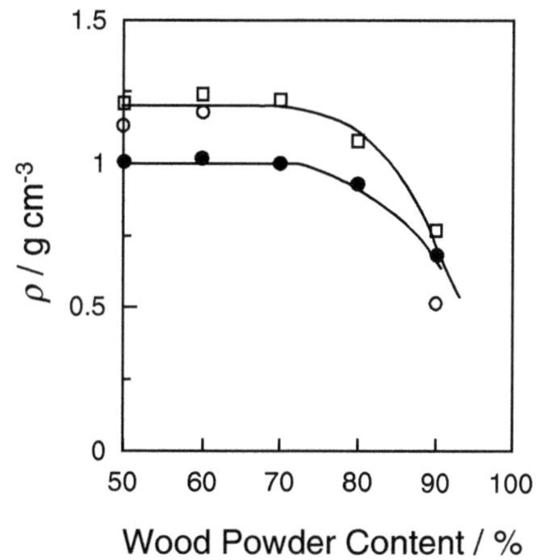

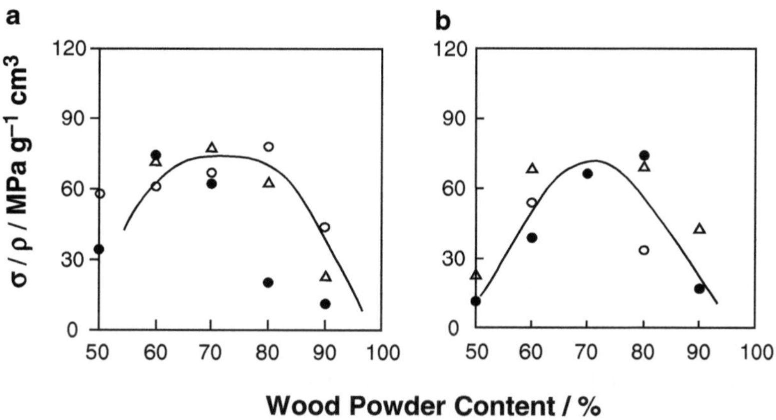

Fig. 40 Relationship between σ/ρ and wood powder content of (**a**) KLD (*open circle*), KLT (*closed circle*) and KLP (*open triangle*) (**b**) LSD (*open circle*), LST (*closed circle*) and LSP (*open triangle*)

molecular mass of ethylene glycol and kinds of lignin are not prominent. The maximum value of $\sigma\rho^{-1}$ was ca. 70 MPa g^{-1} cm^3 for both series of PU composites. The results indicate that densely filled composites having good mechanical performance were obtained.

$E\rho^{-1}$ values of PU composites without veneer plates show a maximum at around 70% of wood powder content. When powder content exceeded 70%, PU composites were delaminated from veneer plates which could be observed visually.

The mechanical properties of commercial composite panels, used for practical applications in the field of civil construction and having the same thickness with the PU panels prepared in this study, were measured using the same method as described in the experimental section. The maximum stress of commercial composite panels was 38 MPa and compression modulus was 2.9 GPa. The obtained maximum stress of the composites prepared in this study ranged from 60 to 70 MPa depending on the kinds of polyols. Accordingly, the maximum stress of composites prepared in this study was 1.5 to 2.0 times. Bending elasticity of the composites was 1.8 times greater than those of commercial panels.

In this study, composites were derived from natural polymers, which can be obtained in a large mass as a residual resource from industries. The mechanical properties showed an optimum point at around 70% filler content. This indicates that composites can be used in a condition where wood powder is densely filled. This is advantageous for biodegradation in nature, since wood powder degrades faster that lignin. From the above results, it can be said that the panels developed in this study are practically usable as construction panels having a biocompatible nature.

6.3.3 PU Composites Filled with Inorganic Fillers

In order to use biocomposites for housing, it is important to prepare thermally stable materials. It is known that, regarding the polymorphic structure of aluminum hydroxide ($Al(OH)_3$), the crystal-crystal transformation occurs by releasing water through the heating process. When $Al(OH)_3$ is mixed with PU derived from lignin, thermal stability, especially inflammability of PU composites markedly improves, since evolved water extinguisgers burning. Various attempts have been made to produce high inflammability by changing preparation condition of composites containing $Al(OH)_3$. Figure 41 shows a simplified preparation scheme to make lignin PU composites filled with $Al(OH)_3$.

Lignin polyols were prepared using KL or LS which was solved in DEG, TEG, or PEG. In order to control the reaction, polyols obtained from ML were also added. Figure 42 shows examples of PU composites derived from LSP/MLP, 50/50 mixture, filled with $Al(OH)_3$. The mixing ratio of $Al(OH)_3$ is shown as wt% in total polyol. The dark brown colour comes from colour of LS.

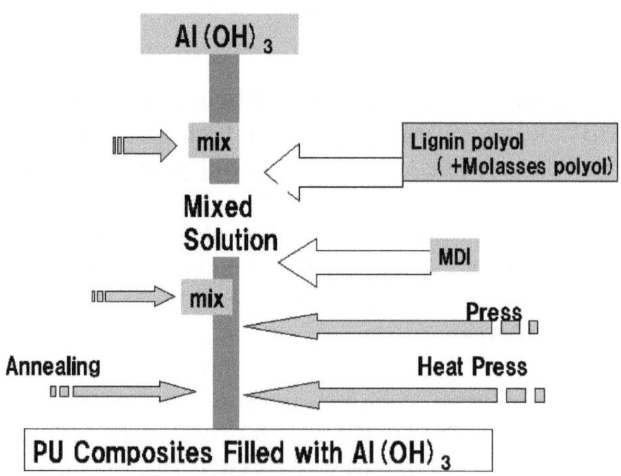

Fig. 41 Preparation scheme of lignin PU composites filled with $Al(OH)_3$

Fig. 42 Examples of PU composites derived from LSP/MLP, 50/50 mixture, filled with Al $(OH)_3$

Fig. 43 Relationships between apparent density, number of oxyethylene chains (n) and Al $(OH)_3$ content in KLP

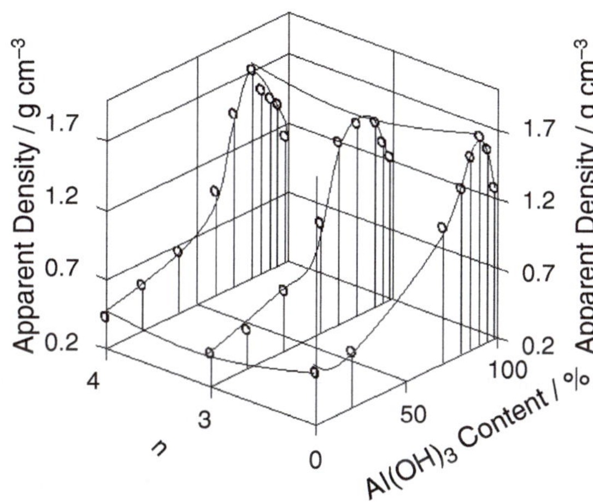

Apparent density of KL PU composites filled with Al $(OH)_3$ is observed in a range 0.4–17 $g\,cm^{-3}$. As shown in Fig. 43, the number of oxyethylene unit (n) scarcely affects apparent density. At the most closely packed condition of Al $(OH)_3$, apparent density shows the maximum value [167]. When lignin polyols were changed from KL to LS, no marked difference was observed. Furthermore, if molasses polyol was mixed with lignin polyol in a broad mixing range in order to control processing, the effect on values of apparent density was in a limited range However, the maximum values of bending strength of the composites slightly increased with increasing lignin polyol content. Figure 44 shows the relationship between bending stress (σ), Al $(OH)_3$ content, and LSP/MLP mixing ratio. It is clearly seen that σ hows the maximum value when Al $(OH)_3$ content is around 80%.

Thermal decomposition of PU composites derived from lignin was investigated by TG. Thermal decomposition occurred in two stages. The high temperature side decomposition observed at around 310–320°C was the major decomposition due

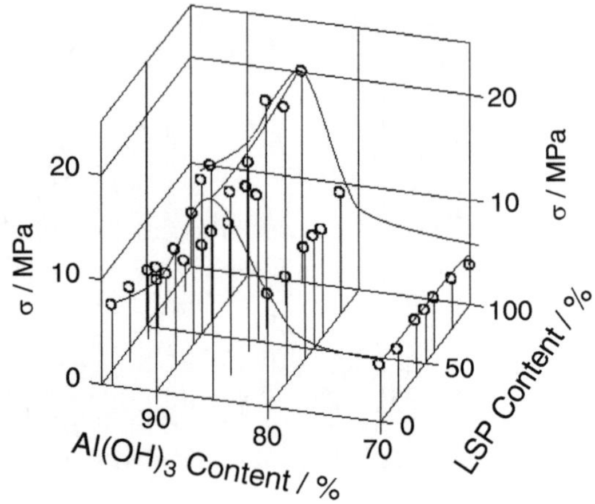

Fig. 44 Relationship between bending stress (σ), Al $(OH)_3$ content and LSP/MLP mixing ratio

to the decomposition of PU. The low temperature side mass change due to water evolution was not prominent when Al $(OH)_3$ mixing ratio was small. However, when combustion was tested according to the standards, both perpendicular and horizontal directions using PU plates, no spreading of fire was observed. The results indicate that PU composites derived from lignin filled with Al $(OH)_3$ are thermally stable.

6.3.4 Geocomposites from Lignin

PU reaction is used in order to stabilize the soil for civil construction, i.e. two kinds of solution were injected directly into soil before tunneling and then the soil was excavated after completion of the reaction. This process is advantageous when removing a large amount of soil using an excavator as it removes the risk of collapse. However, the removed soil cross-linked by synthetic polymers is not biocompatible. In order to solve the above problem, application of lignin polyol to stabilize sand was attempted [168]. The amount of lignin polyol is less than 5% of the total sand and a solidified sand block could be obtained. Compression test of sand blocks was carried out and sufficient strength was confirmed. The conversion from synthetic to biopolymer, such as lignin, in the field of civil construction requires further investigation. It has also attempted to prepare PU composite encapsulated fertilizers that can be maintained for longer time by slow releasing in the soil [169].

6.3.5 Lignin PU Composites Filled with Various Kinds of Wastes from Agriculture, Fishery, Textile, and Food Industries

Lignin PU composites filled with chitin

Chitin is the 2nd major biomaterial produced in the world. As a residue of food industry and fishery, it is a resource which has not been fully utilized. Lignin PU filled with chitin powder was prepared and its physical properties were investigated.

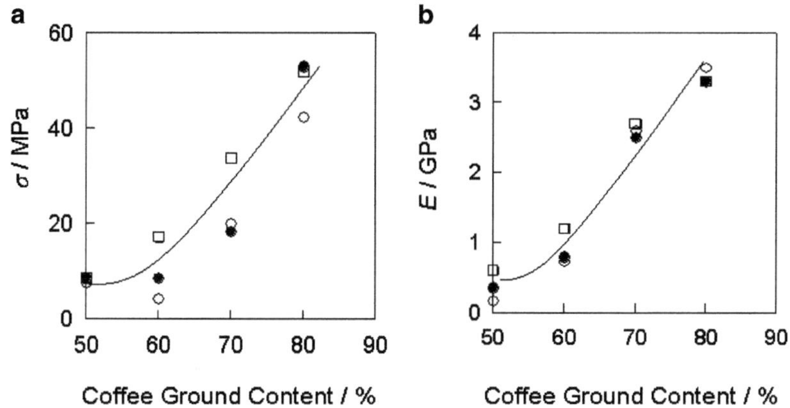

Fig. 45 Relationship between σ, E and coffee ground content. *Closed circle*, KLD; *open circle*, KLT; *open square*, KLP

Characteristics of lignin PU composites filled with chitin are similar to those filled with cellulosic.

Lignin PU composites filled with coffee grounds

Coffee grounds are obtained from large-scale food processing companies as a residue of instant coffee production. They can be obtained independent of season or harvesting time; furthermore, water content is controlled and particle size is even, due to the results of mass production. Coffee parchment can be obtained as a residue of coffee bean plant; however, the quality depends on various factors, and long distance transportation also affects cost performance [170]. PU composites filled with coffee grounds have been investigated for a long time. Here, recent results on lignin PU panel filled with coffee ground are described. Figure 45 shows the relationship between bending stress (σ), bending modulus (E) and coffee ground content in KL polyol solved in DEG, TGE, and PEG. Both σ and E increased linearly with increasing coffee grounds content regardless of kinds of KL polyol. Apparent density of PU composites filled with coffee grounds maintained an almost constant value. On this account, $\sigma\rho^{-1}$, and $E\rho^{-1}$ values also increased with increasing coffee ground content. This indicates that coffee grounds are an important material in future green composites technology.

Lignin PU composites filled with charcoal particles

Carbon particles have been prepared from various kinds of residue of agriculture. Chaff, which contains a large amount of silica, is necessarily produced by rice production. Before synthetic PU foams were widely used, part of the chaff was utilized as a packaging component. At present, chaff is burned at the agriculture site and charcoal mixed with soil. Figure 46 shows lignin PU composites filled with chaff charcoal. As shown in fracture surface of composites, a typical uneven structure of chaff still remains in PU composites. The apparent density of LSPU filled with chaff charcoal 50% in LSD polyol was 0.9 g cm^{-3} and that of chaff charcoal 90% was 0.5 g cm^{-3}. The maximum values (13 MPa) of bending strength and bending

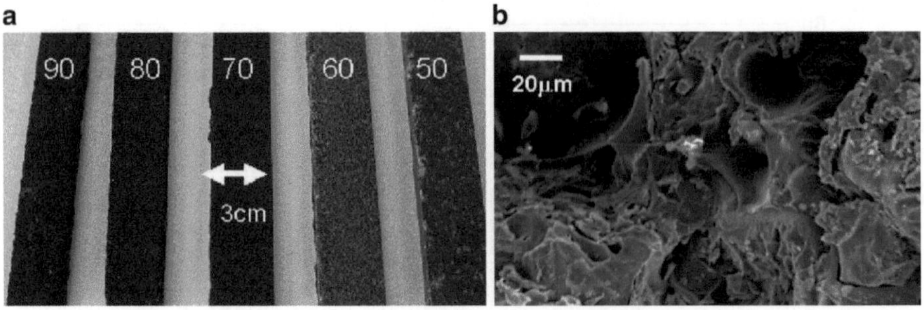

Fig. 46 (**a**) Photographs of LSPU composites filled with various amounts of chaff charcoal. (**b**) Fracture surface of LSPU composites with 90% chaff charcoal. Lignin PU composites filled with waste fabrics

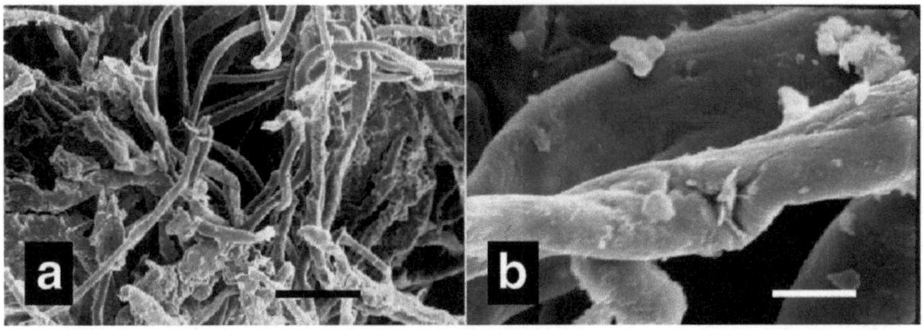

Fig. 47 Scanning electron micrographs LSPU foams filled with waste fabrics. Scale bar (**a**) 100 μm, (**b**) 10 μm

elasticity (2 GPa) were observed at around 60% chaff charcoal content. Thermal decomposition temperature was observed in a temperature from 270 to 285°C.

Lignin PU composites Filled with Waste Fabrics

LS-based PU composites filled with the industrial waste fabrics have been prepared in order to develop insulation board [171]. A large amount of the waste fabrics is obtained in the clothing industries in the cutting process. However, the industrial waste fabrics are discarded without recycling. Figure 47 shows scanning electron micrographs of composites. Waste fabric obtained from the clothing industries consists of various kinds of fabrics. The materials and sizes are in a wide range. Part of a photo (Fig. 47a) is magnified (Fig. 47b). Cellulose fibre is clearly seen. Although the photos are not shown, wool and polyester fibers can be identified. The colour of composites was dark brown, reflecting the colour of lignin polyol Thermal diffusivity of LSP 20% is $2.8 \times 10^{-8}\,m^2\,s^{-1}$, LSP 30% is $4.8 \times 10^{-8}\,m^2\,s^{-1}$, and LSP 40% is $6.5 \times 10^{-8}\,m^2 s^{-1}$ respectively. Thermal diffusivity decreases with increasing filler content. This indicates that fiber wastes used as fillers show a reasonable insulation effect.

6.4 *Epoxy Resins from Lignin*

Epoxy resins are known to be one of the important polymers, since they are used as adhesives, composites matrices, and elastomers. Lignin-based epoxy resins have been investigated [135, 136]. Recently, aliphatic polyester, such as polycaprolactones, poly(ethylene succinate), and poly(lactic acid) have received considerable attention, since they are degradable in natural conditions. In order to develop biocompatible epoxy resins from lignin through various reaction processes, ester-type epoxy resins were prepared from lignin, polyethylene glycol diglycidyl ether, and azelaic anhydride [172]. Other types of epoxy resins were also prepared from ester-carboxylic acid derivatives of AL [173–175].

7 Conclusions

Polymeric features of lignin and its potential as a bio-resource have been reviewed, focusing on its characteristic structure and properties. Industrial lignin obtained in a large scale as a by-product of pulp and biofuel industry can be used "as obtained" for raw materials of various types of polymers, whose physico-chemical properties are superior as natural materials in living tissues that can survive as important constituents of plants in the long term. Lignin, as a representative biopolymer, has potential to compete with synthetic polymers from petroleum in view of both properties and cost performance. It should be considered that lignin will become a promising biocompatible resource in future.

Acknowledgement Authors are grateful to Professor Clive S. Langham for his helpful comments.

References

1. Sarkanen KV, Ludwig CH (1971) Lignins, occurrence, formation, structure and reactions. Wiley, NY
2. Lin SY, Dence CW (1992) Methods in lignin chemistry. Springer, Berlin
3. Hu TQ (2008) Characterization of lignocellulosic materials. Blackwell, Oxford
4. Rezanowich A, Goring DAI (1960) Polyelectrolyte expansion of a lignin sulfonate microgel. J Colloid Sci 15:452–471
5. Ramiah MV, Goring DAI (1965) The thermal expansion of cellulose, hemicelluloses and lignin. J Polym Sci C1:27–48
6. Goring DAI (1971) Polymer properties of lignin and lignin derivatives. In: Sarkanen KV, Ludwig CH (eds) Lignins, occurrence, formation, structure and reactions.Wiley, NY, pp 695–865
7. Donaldson L, Hague J, Snell R (2001) Lignin distribution in coppice poplar, linseed and wheat straw. Holzforschung 55:379–385
8. Barakat A, Chabbert B, Cathala B (2007) Effect of reaction media concentration on the solubility and the chemical structure of lignin model compounds. Phytochemistry 68:118–25
9. Glasser WG, Glasser HR (1974) Simulation of reactions with lignin by computer (SIMREL). I Polymerization of coniferyl alcohol monomers. Macromolecules 7:17–27

10. Glasser WG, Glasser HR (1974) Simulation of reactions with lignin by computer (Simrel) II A model for softwood lignin. Holzforschung 28:5–11
11. Glasser WG, Glasser HR (1976) Simulation of reactions with lignin by computer (Simrel) III The distribution of hydrogen in lignin. Cell Chem Tecnol 10:23–37
12. Adler E, Marton J (1959) Zur Kenntonis der Carbonylgruppen im Lignin I. Acta Chem Scand 13:75–96
13. Nimz HH (1974) Beech lignin-proposal of a constitutional scheme. Angew Chem Int Ed 13:313–321
14. Balakshin MY, Capanema EA, Chang H-M (2008) Recent advances in the isolation and analysis of lignins and lignin-carbohydrate complexes. In: Hu TQ (ed) Characterization of lignocellulosic materials. Blackwell, Oxford, pp 148–170
15. Koshijima T, Watanabe T, Yaku F (1998) Structure and properties of the lignin-carbohydrate polymer as an amphipathick substance. In: Glasser WG, Sarkanen S (eds) Lignin properties and materials. American Chemical Society, Washington, DC, pp11–28
16. Dong D, Fircke AL (1995) Intrinsic viscosity and the molecular weight of kraft lignin. Polym 36:2075–2078
17. Hatakeyama H, Iwashita K, Meshitsuka G Nakano J (1975) Effect of molecular weight on glass transition temperature of lignin. Mokuzai Gakkaishi 21:618–623
18. Mörck R, Yoshida H, Kringstad KP Hatakeyama H (1986) Fractionation of kraft lignin by successive extraction with organic solvents. Functional groups, ^{13}C-NMR spectra and molecular weight distributions. Holzforschung 40(Suppl):51
19. Forss K, Kokkonen R, Sagfors P-E (1989) Detemination of molecular mass distribution studies of lignins by gel permeation chromatography. In: Glasser WG, Sarkanen S (eds) Lignin properties and materials. American Chemical Society, Washington, DC, pp 124–133
20. Gellerstedt G (1992) Gel permeation chromatography. In: Lin SY, Dence CW (eds) Methods in lignin chemistry. Springer, Berlin, pp 487–497
21. Pla F (1992) Light scattering methods. In: Lin SY, Dence CW (eds) Methods in lignin chemistry. Springer, Berlin, pp 498–508
22. Dong D, Fircke AL (1995) Effect of multiple puling variables on the molecular weight and molecular weight distribution of kraft lignin. J Wood Chem Technol 15:369–393
23. Froment P, Pla F (1989) Detemination of average molecular weight and molecular weight distribution of lignin. In: Glasser WG, Sarkanen S (eds) Lignin properties and materials. American Chemical Society, Washington, DC, pp 134–143
24. Pla F (1992) Vapor pressure osmometry. In: Lin SY, Dence CW (eds) Methods in lignin chemistry. Springer, Berlin, pp 509–517
25. Himmel ME, Tatsumoto L, Oh KK Grohmann K, Johnson DK, Chum JL (1989) Determination of a polymer's molecular weight distribution by analytical ultracentrifugation. In: Glasser WG, Sarkanen S (eds) Lignin properties and materials. American Chemical Society, Washington DC, pp 82–99
26. Ben-Ghedalia B, Yosef E (1994) Effect of isolation procedure on molecular weight distribution of wheat straw lignins. J Agric Food Chem 42:649–652
27. Lindner A, Wegner G (1990) Characterization of lignins from organosolv pulping according to the organocell process. Part. 3 Molecular weight determination and investigation of reactions isolated by GPC. J Wood Chem Technol 10:331–350
28. Glennie DW (1971) Reactions in sulfite pulping. In: Sarkanen KV, Ludwig CH (eds) Lignins, occurrence, formation, structure and reactions.Wiley, NY, pp 597–637
29. Rezanowich A, Yean WQ, Goring DAI (1964) High-resolution electron microscopy of sodium ligno sulfonate. J Appl Polym Sci 81:801–1812
30. Lebo SE, Braten SM, Fredheim GE Lutnaes BF, Lauten RA, Myrvold BO, McNally TJ (2008) Recent advances in the characterization of lignosulfonate. In: Hu TQ (ed) Characterization of lignocellulosic materials. Blackwell, Oxford, pp 188–205
31. Glasser WG, Dave V, Frzier CE (1993) Molecular weight distribution of semi-commerical lignin derivatives. J Wood Chem Technol 13:545–559
32. Schiochi EJ, Ward TC, Haney MA, Mahn B (1990) The absolute molecular weight distribution of hydroxypropylated lignins. Macromolecules 23:420–1429

33. Westermark U, Gusafsson K (1994) Molecular size distribution of wood polymers in birch draft pulps. Holzforshung 48:146–150
34. Hatakeyama T, Hatakeyama H (1982) Temperature dependence of x-ray diffractograms of amorphous lignins and polystyrenes. Polymer 23:475–477
35. Hatakeyama T (1982) X-ray analysis of liquid to liquid transition of polystyrene. J Macromol Sci Phys B21:299–305
36. Hatakeyama T, Quinn FX (1999) Thermal analysis, fundamentals and applications to polymer science. 2nd edn, Wiley, UK
37. Hatakeyama T, Hatakeyama H (1995) Effect of chemical structure of amorphous polymers on heat capacity difference at glass transition temperature. Thermochim Acta 267:249–257
38. Yano S, Hatakeyama H, Hatakeyama T (1984) Temperature dependence of the tensile properties of lignin/paper composites. Polymer 25:890–893
39. Hatakeyama H, Kubota K, Nakano J (1972) Thermal analysis of lignin by differential scanning calorimetry. Cell Chem Technol 6:521–529
40. Hatakeyama T, Nakamura K, Hatakeyama H (1982) Studies on heat capacity of cellulose and lignin by differential scanning calorimetry. Polymer 23:1801–1804
41. Hatakeyama H (1992) thermal analysis. In: Lin SY, Dence CW (eds) Wood science, methods in lignin chemistry. Springer, Berlin, pp 200–214
42. Hatakeyama T, Hatakeyama H (2004) Thermal properties of green polymers and biocomposites. Kluwer Academic, Dordrecht
43. Nguyen T, Zavarin E, Barrell EM II (1981) Thermal analysis of lignocellulosic materialas. Part I Unmodified materials. J Macromol Sci Rev Macromol Chem C20:1–65
44. Jain RK, Glasser WG (1993) Lignin derivatives II Functional ethers. Holzforschung 47: 325–332
45. Yano S, Rigdahl M, Kolseth P, Ruvo A (1984) Water-induced softening lignosulfonates Part 1. Effect of molecular mass. Svensk Papperstidn 87:R170–R176
46. Yano S, Rigdahl M, Kolseth P, Ruvo A (1984) Water-induced softening lignosulfonates Part 1. Effect of counter ion. Svensk Papperstidning 88:R10–R14
47. Baumberger S, Dole P, Lapierre C (2002) Using transgenic poplars to elucidate the relationship between the structure and the thermal properties of lingins. J Agric Food Chem 50:2450–2453
48. Laborie M-PG, Salmén L, Frazler CE (2004) Cooperativity analysis of the in situ lignin glass transition. Holzforschung 58:129–133
49. Hatakeyama H, Nakamura K, Hatakeyama T (1980) Studies on factors affecting the molecular motion of lignin and lignin-related polystyrene derivatives. Trans Pulp Pap Canada 81: 105–110
50. Hatakeyama H, Nakano J (1970) Electron spin resonance studies on lignin and lignin model compounds. Cell Chem Technol 4:281–291
51. Hatakeyama H, Nakano J, Hatano A, Migita N (1969) Variation of infrared spectra with temperature for lignin and lignin model compounds. Tappi 52:1724–1728
52. Hatakeyama T, Hirose S, Hatakeyama H (1983) Differential scanning calorimetric studies on bound water in 1.4-dioxane acidolysis lignin. Macromol Chem 184:1265–1274
53. Hatakeyama H, Nakano J (1970) Nuclear magnetic resonance studies on lignin in solid state. Tappi 53:472–475
54. Hatakeyama H, Tsujimoto Y, Hirose S Hatakeyama T (2009) 2009 ISFWPC, OSLO, June 15–18
55. Nakamura K, Hatakeyama T, Hatakeyama H (1981) Differential scanning calorimetric studies on the glass transition temperature of polyhydroxystyrene derivatives containing sorbed water. Polymer 22:473–476
56. Hatakeyama H (1998) In: Hatakeyama T, Liu Z (eds) Handbook of thermal analysis. Wiely, Chichester, pp 144
57. Salmén L (1984) Viscoelastic properties of in situ lignin under water-saturated conditions, J Materials Sci 19:3090–3096
58. Olsson A-M, Salmen L (1997) The effect of lignin composition on the viscoelastic properties of wood. Nordic Pulp and Pap Res J 12:140–144

59. Hatakeyama H, Hatakeyama T (1998) Interaction between water and hydrophilic polymers. Thermochim Acta 308:3–22
60. Hatakeyama T, Nakamura K, Hatakeyama H (1988) Determination of bound water content in polymers by DTA, DSC and TG. Thermochim Acta 123:153–161
61. Hatakeyama H, Hirose S, Hatakeyama T (1989) Differential scanning calorimetry and NMR studies on the water-sodium lignosulfonate system. In: Glasser WG, Sarkanen S (eds) Lignin, properties and materials. ACS Symposium Series 397. ACS, Washington, DC, pp 274–283
62. Woessner DE, Zimmerman J (1963) Nuclear transfer and anisotropic motional spin phenomena: relaxation-time temperature-dependence studies of water adsorbed on silica gel. IV J Chem Phys 67:1590–600
63. Woessner DE, Snowden BS (1970) Pulsed nmr study of water in agar gels. J Colloid Interface Sci 34:290–299
64. Pepper JM, Baylis PET, Adler E (1959) The isolation and properties of lignins obtained by the acidolysis of spruce and aspen woods in dioxane-water medium. Can J Chem 37:1241–248
65. Higuchi T, Tanahshi M, Sato A (1972) Acidolysis of bamboo lignin. I Gas-liquid chromatography and mass spectrometry of acidolysis monomers. Mokuzai Gakkaishi 18:183–189
66. Nakatsubo F, Tanahashi H, Higuchi T (1972) Acidolysis of bamboo lignins II Isolation and identification of acidolysis products. Wood Res 53:9–18
67. Lundquist K (1976) Low-molecular weight lignin hydrolysis products. Appl Polym Symp 28:1393–1407
68. Lapierre C, Roland C, Monties B (1983) Charaterization of poplar lignins acidolysis products capillary gas-liquid chromatography of monomeric compounds. Holzforshung 37:189–198
69. Yu Y, Lou X, Wu H (2008) Some recent advances in hydrolysis of biomass in hot-compressed water and its comparisons with other hydrolysis methods. Energy Fuels 22:46–60
70. Freudenberg K (1939) Uber lignin. Angew Chem 52:362–363
71. Leopold B (1952) Studies of lignin. III Oxidation of wood from *picea abies* (L.) Karst. (Norway spruce) with nitrobenzene and alkali. Acta Chem Scand 6:38–39
72. Creighton RHJ, Gibbs RD, Hibbert H (1944) Studies on lignin and related compounds. LXXV. Alkaline nitrobenzene oxidation of plant materials and application totaxonomic classification. J Am Chem Soc 66:32–37
73. Higuchi T, Ito Y, Shimada M, Kawamura I (1967) Chemical properties of milled wood lignin of grasses. Phytochemisty 6:1551–1556
74. Westermark U (1985) The occurrence of p-hydroxyphenylpropane units in the middle-lamella lignin of spruce (Picea abies). Wood Sci Technol 19:223–232
75. Freudenberg K, Janson A, Knopf E, Haag A (1936) Zur Kenntnis des Lignins. Chem Ber 69:1415–1425
76. Larsson S, Miksche GE (1967) Gas chromatographic analysis of lignin oxidation products. The diphenyl ether linkage in lignin. Acta Chem Scand 21:1970–1971
77. Glasser WG, Morohoshi N (1979) The structure of lignin in pulps. The distribution of interunit linkages. Tappi 62:101–105
78. Morohoshi N, Glasser WG (1979) The structure of lignins in pulps. Part 5 Gas and gel permeation chromatography of permanganate oxidation products. Wood Sci Technol 13:249–264
79. Gellerstedt G, Lindfors E (1984) Structural changes in lignin during kraft pulping. Holzforschung 38:151–158
80. Gellersted G, Northey RA (1989) Analysis of birch wood by oxidative degradation. Wood Sci Technol 23:75–83
81. Langvik VA, Aakerback NH, Bjarne D (1994) Characterization of aromatic structures in humic and fulvic acids. Environ Int 20:61–5
82. Tamminen T, Poppius-Levlin K, Aurela B, Hortling B (1997) Oxidative degradation of lignin by potassium permanganate. Method improvement. Holzforschung 51:155–157
83. Sakakibara A (1992) hydrogenolysis. In: Lin SY, Dence CW eds. Wood science, methods in lignin chemistry, pp. 350–368. Springer, Berlin
84. Sarkanen KV, Islam A, Anderson CD (1992) Ozonation. In: Lin SY, Dence CW (eds) Wood science, methods in lignin chemistry. Springer, Berlin, pp 387–406

85. Hatakeyama H, Tonooka T, Nakano J, Migita N (1967) Ozonization of lignin model compounds. Kogyo Kagaku Zasshi. 70:2348–2352
86. Tsutsumi Y, Islam A, Anderson D, Sarkanen KV (1990) Acidic permanganate oxidation of lignin model compounds: comparison with ozonolysis. Holzforschung 44:59–66
87. Miksche GE (1972) Zum alkalischen Abbau der p-Alkoxyarylglycerin-β-arylatherstrukuturen des Lignins. Versuchemit erythro–Veratorylglycerin- β-guaiacylather. Acta Chem Scand 26:3275–3281
88. Mbachu RAD, Manley RSJ (2003) Degradaton of lignin by ozone. I The kinetics of lignin degradation of ozone. J Polym Sci Chem Ed 19:2053–2063
89. Letumier F, Barbe J-M, Trichet A, Guilard R (1999) Ozonation reactions of monomer and dimer lignin models: influence of a catalytic amount of a manganese cyclam derivaties on the ozonation reaction. Ozone: Science & Engineering – The Journal of the International Ozone Association 21:53–67
90. Widsten P, Hortling B, Poppius-Levlin K (2004) Ozonation of conventional kraft and superbatch residual lignins in methanol/water and water. Holzforschung 58:363–368
91. Mbachu RAD, Manley RSJ (1981) Degradation of lignin by ozone. II. Molecular weights and molecular weight distributions of the alkali-soluble degradation products. J Polym Sci, Polym Chem Edn 19:2065–78
92. Yokoyama C, Nishi K, Takahashi S (1994) Functional group analysis of reaction products from thermolysis of lignin in ethylene glycol and diethylene glycol. Sekiyu Gakkaishi 37:576–583
93. Yokoyama C, Nishi K, Nakjima A Kazue S (1998) Thermolysis of organosolv lignin. I Supercritical water and supercritical methanol. Sekiyu Gakkaishi 41:243–250
94. Simonova VV (2008) Thermolysis and steam activation of a mixture of lignin and washed-off petroleum water in the presence of copper acetate. Solid Fuel Chem 42:49–53
95. Wahyudiono, Kanetake T, Sasaki M, Goto M (2007) Decomposition of a lignin model compound under hydrothermal conditions. Chem Eng Technol 30:1113–1122
96. Klein MT, Virk PS (2008) Modeling of lignin thermolysis. Energy Fuels 22:2175–2182
97. Hatakeyama T, Nakamura K, Hatakeyama H (1978) Differential thermal analysis of styrene derivatives related to lignin. Polymer 19:593–594
98. Hirose S, Hatakeyama H (1986) A kinetic study on lignin pyrolysis using integral method. Mokuzai Gakkishi 32:621–625
99. Jakob E, Faix O, Till F, Szénely TJ (1995) Thermogramivery/mass spectrometry study of six lignins within the scope of an international round robin test. Anal Appl Pyrolysis 35:167–179
100. Serio MA, Charpenay S, Bassilakins R Solomon PR (1994) Measurement and modeling of lingin pyrolysis. Biomass Bioenergy 7:107–124
101. Hirose S, Kobashigawa K, Izuta Y, Hatakeyama H (1998) Thermal degradation of polyurethanes containing lignin studied by TG-FTIR. Polym Int 47:247–256
102. Caballero JA, Font R, Marcilla A (1997) Pyrolysis of kraft lignin: yields and correlations. J Anal Appl Pyrolysis 39:161–183
103. Fahmi R, Bridgwater A, Thain SC, Donnison IS, Morris PM, Yates N (2007) Prediction of Klason lignin and lignin thermal degradation products by Py-GC/MS in a collection of Lolium and Festuca grasses. J Anal Appl Pyrol 80:16–23
104. Yang H, Yan R, Chen H, Zheng C, Lee DH, Liang DT (2006) In-depth investigation of biomass pyrolysis based on three major components: hemicellulose, cellulose and lignin. Energy Fuels 20:388–393
105. Hatakeyama T, Nakamura K, Hatakeyama H (2000) Vaporization of bound water associated with cellulose fibres. Thermochim Acta 352–353:233–239
106. Lindberg JJ, Kuusela TA, Levon K (1989) Specialty polymers from lignin. In: Glasser WG, Sarkanen S (eds) Lignin properties and materials, ACS Symposium Series 397. American Chemical Society, Washington, DC, pp 134–143
107. Hatakeyama H, Hirose S, Hatakeyama T (1989) High-performance polymers from lignin degradation products. In: Glasser WG, Sarkanen S (eds) Lignin, properties and materials. ACS Symposium Series 397. American Chemical Society, Washington, DC, pp 205–218

108. Hatakeyama H, Hirose S, Yano S, Nakamura K, Hatakeyama T (1990) High-performance materials from lignocellulose. In: Kennedy JF, Phyllips GO, Williams PA (eds) Cellulose sources and exploitation. Ellis Horwood, Chichester, pp 466–471
109. Hatakeyama H, Hayashi E, Haraguchi T (1977) Biodegradation of poly(3-methoxy-4-hydroxy styrene). Polym J 18:759–763
110. Nakamura K, Hatakeyama T, Hatakeyama H (1983) Effect of substituent groups on hydrogen bonding of polyhydroxystyrene derivatives. Polym J 15:361–366
111. Nakamura K, Hatakeyama T, Hatakeyama H (1986) DSC Studies on hydrogen bonding of poly(4-hydroxy-3, 5-dimethoxystyrene) and related derivatives. Polym J 18:219–225
112. Nakamura K, Hatakeyama T, Hatakeyama H (1983) 2 Relationship between hydrogen bonding and bound water in polyhydroxystyrene derivatives. Polymer 24:871–876
113. Hirose S, Nakamura K, Hatakeyama T (1989) Molecular design of linear aromatic polymers derived from phenols related to lignin. In: Schuerch C (ed) Cellulose and wood. Wiley, NY, pp 1133–1144
114. Hirose S, Hatakeyama T, Hatakeyama H (1978) Synthesis and thermal analysis of polyethers related to lignin. Cell Chem Technol 12:713–720
115. Hatakeyama T, Yoshida H, Hirose S, Hatakeyama H (1990) Enthalpy relaxation of polyethers having phenylene groups in the main chain. Thermochim Acta 163:175–182
116. Hirose S, Yoshida H, Hatakeyama T, Hatakeyama H (1992) Thermal behavior of aromatic poymers derived from penols related to lignin. In: Glasser WG, Hatakeyama H (eds) Viscoelasticity of biomaterials, ACS Symposium Series 489. ACS, Washington, DC, pp 385–394
117. Hirose S, Hatakeyama H, Hatakeyama T (1983) Synthesis and thermal analysis of polyacylhydrazones having guaiacyl units with alkylene groups. Sen-i Gakkaishi 39:T496–T500
118. Hirose S, Hatakeyama T, Hatakeyama H (1982) Synthesis and thermal analysis of polyesters having spiro-dioxane rings and guaiacyl units. Sen-i Gakkaishi 38:T507–T511
119. Hirose S, Hatakeyama H, Hatakeyama T (1986) Isothermal crystallization of polybenzalazine derivatives synthesized from vanillin. Sen-i Gakkaishi 42:T49–T53
120. Hirose S, Hatakeyama H, Hatakeyama T (1989) Synthesis and thermal analysis of aromatic polyesters derived from Phenols related to lignin. Wood Process Util 22:181–186
121. Hirose S, Hatakeyama T, Hatakeyama H (1993) Synthesis and thermal analysis of aromatic polyethers derived from degradation products of lignin. In: Kennedy JF, Phillips GO, Williams PA (eds) Cellulosic:chemical, biochemical and material aspects. Ellis Horwood, Chichester, pp 381–387
122. Belgacem MN, Blayo A, Gandini A (1996) Surface characterization of polysaccharides, lignins, printing ink pigments, and ink fillers by inverse gas chromatography. J Colloid Interface Sci 182:431–436
123. Belgacem MN, Blayo A, Gandini A (2003) Organosolv lignin as a filler in inks, varnishes and paints. Indust Crop Produc 18:145–153
124. Ghosh I, Jain RK, Glasser WG (1999) Multiphase materials with lignin. Part 15. Blends of cellulose acetate butyrate with lignin esters. J Appl Polym Sci 74:448–457
125. Ghosh I, Jain RK, Glasser WG (2000) Multiphase materials with lignin. Part 16. Blends of biodegradable thermoplastics with lignin esters. In: Lignin: historical, biological, and materials perspectives, ACS Symposium Series 742. American Chemical Society, Washington DC, pp 331–350
126. Dave V, Glasser WG (1997) Cellulose-based fibers from liquid crystalline solutions: 5 Processing and morphology of CAB blends with lignin. Polymer 38:2121–2126
127. Glasser WG, Wang HX (1989) Derivatives of lignin and ligninlike models with acrylate functionality. In: Glasser WG, Sarkanen S (eds) Lignin: properties and materials Symp Series 397. American Chemical Society, Washington, DC, pp 515–22
128. Glasser WG, Jain RK (1993) Lignin derivatives. I Alkanoates Holzforschung 47:225–33
129. Kelley SS, Glasser WG, Ward TC (1988) Engineering plastics from lignin. XIV Characterization of chain-extended hydroxypropyl lignins. J Wood Chem Technol 8:341–59
130. Rials TG, Glasser WG (1990) Multiphase materials with lignin: 5 Effect of lignin structure on hydroxypropyl cellulose blend morphology. Polymer 31:1333–1338

131. Ciemniecki SL, Glasser WG (1989) Polymer blends with hydroxypropyl lignin. In: Glasser WG, Sarkanen S (eds) Lignin properties and materials. Symposium Series 397. American Chemical Society, Washington, DC, pp 452–63
132. Rials TG, Glasser WG (1989) Multiphase materials with lignin. IV Blends of hydroxypropyl cellulose with lignin. J Appl Polym Sci 37:2399–415
133. Demaret V, Glasser WG (1989) Multiphase materials with lignin. 7 Block copolymers from hydroxypropyl lignin and cellulose triacetate. Polymer 30:570–575
134. Kelley SS, Ward TC, Glasser WG (1990) Multiphase materials with lignin. VIII. Interpenetrating polymer networks from polyurethanes and poly(methyl methacrylate). J Appl Polym Sci 41:2813–28
135. Hofmann K, Glasser WG (1993) Engineering plastics from lignin. 21. Synthesis and properties of epoxidized lignin-poly(propylene oxide) copolymers. J Wood Chem Technol 13:73–95
136. Hofmann K, Glasser WG (1993) Engineering plastics from lignin. 22. Cure of lignin-based epoxy resins. J Adhes 40:229–241
137. Takemura A, Glasser WG (1993) Multiphase materials with lignin. 10. A novel graft copolymer with hydroxyalkyl lignin. Mokuzai Gakkaishi 39:198–205
138. De Oliveira W, Glasser WG (1993) Comparison of some molecular characteristics of starblock copolymers with lignin. In: Kennedy JF, Phillips GO, Williams PA (eds) Cell Chem Biochem Mater Aspects. Horwood, London, UK, pp 263–271
139. De Oliveira W, Glasser WG (1994) Multiphase materials with lignin. 11. Starlike copolymers with caprolactone. Macromolecules 27:5–11
140. De Oliveira W, Glasser WG (1994) Multiphase materials with lignin 13. Block copolymers with cellulose propionate. Polymer 35:1977–1985
141. De Oliveira W, Glasser WG (1994) Multiphase materials with lignin. 14. Star-like copolymers with styrene. J Wood Chem Technol 14:119–126
142. De Oliveira W, Glasser WG (1994) Multiphase materials with lignin 7 Blends of poly(vinyl chloride) with lignin-caprolactone copolymers. J Appl Polym Sci 51:563–571
143. Toffey A, Glasser WG (1997) Cure characterization of polyurethanes with lignin and cellulose derivatives. Holzforschung 51:71–78
144. Gandini A, Naceur BM, Guo ZX, Montanari S (2002) Lignins as macromonomers for polyesters and polyurethanes. In: Hu TQ (eds) Chemical modification, properties, and usage of lignin. Academic/Plenum, NY, pp 57–80
145. Mikame K, Funaoka M (2006) Polymer structure of lignophenol II-comparison of molecular morphology of lignophenol and conventional lignins. Polym J 38:592–596
146. Kimura T, Terada M, Tamura H, Funaoka M (2007) Relationships between the molecular structure of lignins and the performance of the negative plates of lead-acid batteries. Trans Mat Res Soc Jap 32:1091–1094
147. Khunathai K, Parajuli D, Adhikari CR, Kawakita H, Ohto K, Inoue K, Funaoka M (2007) Selective recovery of precious metals by a novel lignin gel. J Ion Exch 8:498–501
148. Parajuli D, Kawakita H, Inoue K, Funaoka M (2006) Recovery of gold (III), palladium(II), and platinum(IV) by aminated lignin derivatives. Indust Eng Chem Res 45:6405–6412
149. Parajuli D, Inoue K, Ohto K, Oshima T, Murota A, Funaoka M, Makino K (2005) Adsorption of heavy metals on crosslinked lignocatechol: a modified lignin gel. React Func Polym 62:129–139
150. Aoyagi M, Funaoka M (2004) A new polymeric photosensitizer for dye-sensitized solar cell with porous TiO_2 from forest carbon resources. J Photochem Photobiol A Chem 164:53–60
151. Saunders J, Fisch K (1962) Polyurethanes. Chemistry and technology in high polymers XV. Interscience, NY
152. Planck H, Egbers G, Syré I (1984) Polyurethanes in biomedical engineering. Elsevier, Amsterdam
153. Saraf VP, Glasser WG, Wilkes G (1985) Engineering plastics from lignin. VII. Structure property relationships of polybutadiene glycol-containing polyurethane networks. J Appl Polym Sci 30:3809–3823

154. Nakamura K, Morck R, Reimann A, Kringstad KP, Hatakeyama H (1991) Mechanical properties of solvolysis lignin derived polyurethanes. Polym Adv Technol 2:41–47
155. Nakamura K, Hatakeyama T, Hatakeyama H (1992) Thermal properties of solvolysis lignin-derived polyurethanes. Polym Adv Technol 3:151–155
156. Hatakeyema H (2002) Thermal analysis of environmentally compatible polymers containing plant components in the main chain. J Therm Anal Calorim 70:755–955
157. Hatakeyama T, Matsumoto Y, Asano Y, Hatakeyama (2004) H Glass transition of rigid polyurethane foams derived from sodium lignosulfonate mixed with diethylene, triethylene and polyethylene glycols. Thermochim Acta 416:29–33
158. Zetterlund P, Hirose S, Hatakeyama T, Hatakeyama H, Albertsson A-C (1997) Thermal and Mechanical properties of polyurethanes derived from mono- and disaccharides. Polym Int 42:1–8
159. Hatakeyama H, Nakamura K, Kobashigawa K, Morohoshi N (1995) Biodegradable polyurethanes from plant components. J Macromol Sci A32:743–750
160. Hatakeyama H, Hatakeyama T (2005) Environmentally compatible hybrid-type polyurethane foams containing saccharide and lignin components. Macromol Symp 224:219–226
161. Hatakeyama H, Kosugi R, Hatakeyama T (2008) Thermal properties of lignin- and molasses-based polyurethane foams. J Therm Anal Calorim 92:419–424
162. Albertsson AC, Varma IK (2002) Degradation of aliphatic polyesters. Adv Polym Sci 157: 1–179
163. Hatakeyama H, Yoshida T, Hirose S, Hatakeyama T (2000) Thermal and visoelastic properties of cellulose-and lignin-based polycarolactones. In: Kennedy JF, Phillips GO, Williams PA, Lönnberg B (eds) Cellulosic pulps, fibres and materials. Woodhead, Cambridge, pp 327–336
164. Hatakeyama T, Izuta Y, Hirose S, Hatakeyama H (2002) Phase transitions of lignin-based polycaprolactones and their polyurethane derivatives. Polymer 43:177–1182
165. Hatakeyama H, Tanamachi N, Matsumura H, Hirose S, Hatakeyama T (2005) Biobased polyurethane composite foams with inorganic fillers studied by thermogravimetry. Thermochim Acta 431:155–160
166. Hatakeyama H, Nanbo T, Hatakeyama T (2008–1) Thermal and mechanical analysis of lignocellulose-based biocomposites. In: Hu TQ (ed) Characterization of lingocellulosic materials. Blackwell, UK, pp 275–287
167. Funabashi M, Hirose S, Hatakeyama T, Hatakeyama H (2003) Effect of filler shape on mechanical properties of rigid polyurethane composites containing plant particles. Macromol Symp. 197:231–241
168. Hatakeyama H, Nakayachi, Hatakeyama T (2005) Thermal and mechanical properties of polyurethane-based geocomposites derived from lignin and molasses. Comp Part A Appl Sci Manuf 36:698–704
169. Yamauchi N, Hirose S, Hatakeyama H (2001) Preparation and thermal properties of polyurethane composite containing fertilizer. In: Kennedy JF, Phillips GO, Williams PA, Hatakeyama H (eds) Recent advances in environmentally compatible polymers. Woodhead, Cambridge, pp 211–216
170. Hatakeyama H, Kamakura D, Kasahara H, Hirose S, Hatakeyama T (2001) Biodegradable polyurethane composite containing coffee bean parchments. In: Kennedy JF, Phillips GO, Williams PA, Hatakeyama H (eds) Recent advances in environmentally compatible polymers. Woodhead Pub, Cambridge, pp 191–196
171. Tanamaich N, Hatakeyama H, Funabashi M, Hatakeyama T (2007) Thermal diffuisitivity of polyurethane composites filled with industrial waste fabrics. Sen-i Gakkaishi 63:182–184
172. Hirose S, Kobayashi M, Kimura H, Hatakeyama H (2001) Synthesis of lignin-based polyester-epoxy resins. In: Kennedy JF, Phillips GO, Williams PA, Hatakeyama H (eds) Recent advances in environmentally compatible polymers. Woodhead, Cambridge, pp 73–78
173. Hirose S, Hatakeyama T, Hatakeyama H (2003) Synthesis and thermal properties of epoxy resins from ester- carboxylic acid derivative of alcoholysis lignin. Macromol Symp 197: 157–169

174. Hirose S, Hatakeyama T, Hatakeyama H (2005) Curing and glass transition of epoxy resins from ester-carboxylic acid derivatives of mono- and disaccharides and alcoholysis lignin. Macromol Symp 224:343–353
175. Hirose S, Hatakeyama T, Hatakeyama H (2005) Glass transition and thermal decomposition of epoxy resins from the carboxylic acid system consisting of ester-carboxylic acid derivatives of alcoholysis lignin and ethylene glycol with various eicarboxylic acids. Thermochim Acta 431:76–80

Editor: S. Kobayashi

Adv Polym Sci (2010) 232: 65–96
DOI:10.1007/12_2009_41

Published online: 18 February 2010

Tensile Mechanics of α-Helical Coil Springs

Atsushi Ikai

Abstract This chapter reviews the mechanical properties of the molecular-size coil spring, i.e., α-helix of polypeptides. Since the geometry of α-helix was established both by the model based prediction of Pauling and his colleagues and by the X-ray crystallography of proteins, its fundamental importance in the building process of secondary and tertiary structures of proteins has been widely recognized. Furthermore, the relative stability of α-helix against non-helical states has been rigorously treated by statistical mechanics with experimental verifications on model polypeptides and real proteins. Its mechanical properties including its elastic modulus have, however, been discussed mainly from theoretical point of view due to the paucity of experimental data, especially those obtained at the single-molecule level. Unlike other properties of protein molecules, the mechanical properties must be measured at the single molecular level. A recent development of such experimental techniques based mainly on atomic force microscopy now provides us access to single molecular level data and enables us to discuss the mechanical properties of polypeptides and proteins based on various theoretical frameworks that were previously developed and concurrently being investigated. In this review, recent observations from the single-molecule-stretching experiments of α-helical polypeptides are summarized and discussed in consideration of recent theoretical developments.

Keywords α-Helical coil spring · Atomic force microscopy · Force spectroscopy · Mechanical unfolding of α-helix · Peptide hydrogen bond · Tensile mechanics of α-helix · Young's modulus

---

A. Ikai (✉)
Innovation Laboratory, Tokyo Institute of Technology, 4259 Nagatsuta, Midori-ku, Yokohama 226-8501, Japan
e-mail: ikai.a.aa@m.titech.ac.jp

Contents

1 Introduction

As bottom up nanotechnology is gaining momentum year by year, precise characterizations of nanometer sized synthetic as well as natural materials at the molecular level have been gaining increasing importance. In synthetic science fields, one of the vigorous trends has been biomimetic chemistry where natural materials with challenging structures and advantageous properties to human life were intensely examined for the discovery of new ways to reproduce the best of their qualities one way or another. Based on the results of such research, high quality synthetic molecules and materials have been produced [1]. Examples of such natural materials that would make our life immensely more comfortable are numerous and abundantly displayed [2]. An important concept here is to improve the macroscopic material properties based on the knowledge obtained from the analysis of microscopic geometries, chemical compositions, and molecular level properties of materials found in nature. In today's nano-technology, where the use of single molecules as components of an assembled device is vigorously tried, the elucidation of the physical and mechanical properties of candidate molecules at the single molecular level is becoming more and more important [3–7]. When a new molecular machine of useful function is assembled, the performance of individual molecules as independent parts would be a crucial factor [8]. Examples of such molecular machines found in nature are most often composed of bio-macromolecules such as proteins, nucleic acids, lipids and possibly carbohydrates and driven by the storage of chemical potential in one form or another. Since most of the bio-macromolecules mentioned above are electrical insulators, information transfer from one component molecule to another is done mainly through mechanical means. Force transduction, therefore, is an important subject to be pursued in the study of biological systems [9–11]. There are abundant examples where tiny forces are acting to direct the life processes to proceed in due coupling with biochemical reactions. Flagella, as one example, promote the translation motion of bacteria in liquid media [12] and cytoskeletal structures, as another example, keep the shape of the cell and provide the basic force for the cell to move in conjunction with force regulated biochemical reactions [13].

We are here, in this short review, interested in the mechanical behavior of a single polypeptide chain, especially when it is in helical conformations because there are many cases where it is acting as an important structural and functional element of protein based bio-structures and would provide a molecular-size coil springs as an important molecular device in future nano-machines.

The field of macroscopic biomechanics has been established as a discipline within bioengineering [14] but its extension to nanoscience and nanotechnology is relatively recent. Since the means of biological information and motion transfer at the molecular level is basically mechanical, the extension of biomechanics to nano-biomechanics is a reasonable development both from theoretical as well as experimental points of view.

Measurement of mechanical properties of proteins, especially those of fibrous proteins, has been an important interdisciplinary concern in the history of protein science. In fact, the very early X-ray work by Astbury and his colleagues established the force dependent conformational transition of keratin fiber between α- and β-forms [15]. A large body of work has since been accumulated on the measurement of mechanical parameters of fibrous structures made of keratin, collagen, dentin and other structural proteins [10, 14, 16, 17]. Measurement was done at the macroscopic level on higher order assemblies of fibrous proteins, applying established methods in materials science for the determination of, for example, static and/or dynamic elastic modulus [14].

The measurement of bulk modulus of macroscopic fibers provided the fundamental knowledge to understand the physical properties of protein based structures, such as wool, tendon, bone, connective tissues, silk, etc. in relation to their chemical composition and eventually opened a variety of ways to improve their advantageous properties or mimic them in partially or totally synthetic materials [18].

In basic science, as our understanding of the mechanism of muscle contraction is improved, measurement of mechanical properties of component proteins became an important issue and the elastic modulus of a single actin fiber was measured [19] using a tapered glass rod as a force transducer. The work was influential as one of the earliest microscopic measurements of mechanical properties of protein based structures. When the Young's modulus of a single filament was translated to that of single molecule of actin, a value of 2 GPa was obtained. Similar work on an actin based thin filament, as isolated from detergent solubilized cell components, was performed, giving numerical estimates of the non-linear Young's modulus and the tensile strength [20]. The non-linear increase of the apparent Young's modulus of biological materials as a function of relative strain up to the breakpoint of the sample is now a well recognized observation as an example of strain stiffening or strain hardening. It is, however, in contrast to the stress-strain behavior of synthetic polymer materials which more often shows an initial linear increase of the tensile force against extension followed by a gradual flattening of the curve before reaching maximum tensile strength [14]. In general, the mechanics of deformation, fatigue, and fracture of bio-macromolecules and biomaterials is an emerging discipline waiting for active participation of both biologists and engineers. It will be advanced in accordance with the theoretical and experimental methods developed in mechanical engineering [21]. For example, the tensile properties of biological structures is beginning to be treated in terms of a 2D or 3D network theory but the molecular details of individual cases are not clarified yet [9].

2 Recent Technical Developments to Promote Nano-Biomechanics

The advent of nano-mechanical devices and associated technologies in terms of atomic force microscopy (AFM) and/or optical tweezers has led to the development of more advanced single molecular mechanical measurement and manipulation methods [11]. Florin et al. showed that AFM could be used to measure the unbinding force of ligand-receptor pair at the single molecular level [22]. Mitsui et al. reported the result of application of tensile force to a single molecule of globular protein reporting numerical values of molecular stiffness [23]. Rief et al. showed a regular sawtooth pattern obtained on mechanical unraveling of tandemly connected globular subunits in a giant molecule of titin which has since then become a standard method of stretching proteins [24]. Alam et al. [25] and Hertadi et al. [26, 27] presented mechanical unfolding of globular proteins at the single molecular level. A large body of work on mechanical unfolding of proteins has since then been presented [28]. An excellent review on the methodology of cantilever based force measurement is given in [29]. Briefly, in the force mode of AFM, a micro-fabricated cantilever is used as a mechanical force sensor with a tiny probe in the form of a stylus at its free end. Any force acting on the stylus, either normally or laterally, bends or twists the cantilever and the deflection, d, of the cantilever is accurately detected by, for example, an optical lever system. By knowing the force constant, k, of the cantilever, the magnitude of the force, f, is calculated as $f = k \times d$. Extension (or compression) of the sample, E, under the application of this force is obtained as the difference between the piezo movement, D, and the deflection of the cantilever as $E = D - d$. The situation is schematically depicted in Fig. 1a as an image of experimental setup and, in Fig. 1b, definitions of D, d (positive for downward and negative for upward deflection) and E are given.

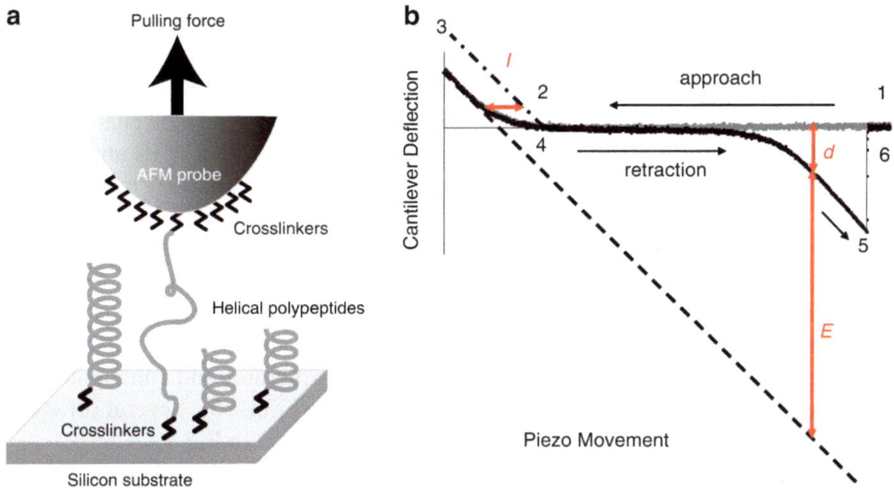

Fig. 1 Force mode of AFM. (**a**) A schematic view of single molecule stretching. (**b**) Relation between piezo distance, D, cantilever deflection, d, and sample extension, E. The approach of the cantilever starts from position 1 on the right to position 2 where the probe touches the sample surface. When the sample surface is rigid, the cantilever is pushed up to position 3 where the cantilever movement is reversed and the cantilever traces back to position 1 without hysteresis. When the sample surface is soft, cantilever deflection follows a curve from position 2 to 3' indenting the sample with the maximum depth of I. In the retraction regime, when a part of the sample is adhered to the probe, the cantilever shows a gradual or immediate downward deflection due to the tensile force from the sample. In this figure, the tensile material adhered to the probe is assumed to be a flexible polymer like material so that the downward deflection of the cantilever is initially small but rapidly increases to position 5 where the adhesion bond of the sample to the probe is broken abruptly (Reproduced from [66] with permission)

3 The α-Helix

The α-helix looks like a coil spring at the molecular level but we have not enough knowledge about its mechanical properties to answer such questions as follows. Is it useful as a component in molecular devices to be developed in the future? Is it robust enough to support many force dependent processes in native condition of life processes?

The α-helix is one of the two major motifs of the secondary structure of proteins, the other being the β-sheet. The names were derived after the finding of Astbury on two alternative conformations of wool keratin on the basis of X-ray fiber diagram [14]. He found that moist wool keratin showed a transition from a coiled (meaning tangled) conformation to a more extended form on the application of a tensile force. He called the coiled conformation α-form and the extended one β-form and proposed a helical conformation for the α-form. The α-form was characterized by a strong axial diffraction for a 0.51-nm repeating structure. After several proposals for the conformations of α- and β-forms, Pauling presented his classical model of α-helix and β-sheet together with Corey and Branson [30]. Their enormously successful model as later verified was based on a non-integral number of amino acid residues, i.e., 3.6 residues per 1 helical turn and a maximized number of hydrogen bonds between >NH and >C=O of the main polypeptide chain. When the crystal structure of myoglobin, a protein with more than 70% of its amino acid residues in α-helical conformation, was solved by Kendrew et al. [31] revealing its detailed atomic arrangement, it was clear that what Pauling et al. proposed was essentially correct. Since the first example of myoglobin, it has become customary to classify proteins by the topographic arrangements of α-helices, β-sheets, and other secondary structural motifs [32]. Thermodynamic stability and mechanical rigidity of a protein molecule along with its functional integrity mostly do, therefore, stem from basic properties of α-helix and β-sheet. The structural and functional significance of α-helix in proteins is still widely discussed [33–35]. Several different synthetic polypeptides to be treated in this chapter are given in Fig. 2.

In the basic geometry of the right handed α-helix, the main chain forms a regular turns of helix with a pitch of 0.54 nm and 3.6 amino acid residues per turn corresponding to 100° rotation per peptide group around the helical axis. The per residue translation along the helical axis is 0.15 nm [30]. The helical pitch of an α-helix in coiled coil conformation becomes 0.51 nm as observed for keratin by Astbury. Side chains are all outside of the helix. The >NH group of every peptide group forms a hydrogen bond with the >C=O group of the amino acid four residues ahead ($i \rightarrow i+4$ hydrogen bonds). A similar helical structure called 3_{10} helix which contains an integral number of three residues per turn forms hydrogen bonds between $i \rightarrow i+3$ residues.

Dihedral angles, ϕ and ψ, are defined as shown in Fig. 3.

Since, out of three possible dihedral angles of the main chain peptide group, the −NH–C(=O)− bond is normally fixed to a planar trans configuration, the corresponding dihedral angle, ω, being fixed at 180°, the ϕ, ψ dihedral angles alone determine the conformation of the peptide groups along the main chain.

Due to the partial double bond character of the amide bonds, the distance connecting consecutive α-carbon atoms remains constant, being unaffected by the internal bond rotations. Virtual bonds C^{α}–C^{α} defined in this manner are often adopted in treating the configurational characteristics of peptide chains.

Ramachandran and his colleagues made this point clear by representing the conformational states of peptide group on a 2D map of ϕ, ψ, which has since then been called the Ramachandran plot [36, 37]. By simply excluding high energy regions in the Ramachandran plot where overlapping of atoms occurs causing steric repulsions, low energy conformational states for any peptide can be mapped in rather limited regions.

Residues in α-helices typically adopt backbone (ϕ, ψ) dihedral angles around (−60°, −45°), the sum of the ψ dihedral angle of one residue and the ϕ dihedral angle of the next residue being approximately −105°. Consequently, dihedral angles of α-helix fall on a diagonal region on the Ramachandran plot ranging approximately from (−90°, −15°) to (−35°, −70°). Figure 4 shows a Ramachandran plot of an alanine residue in poly-L-alanine.

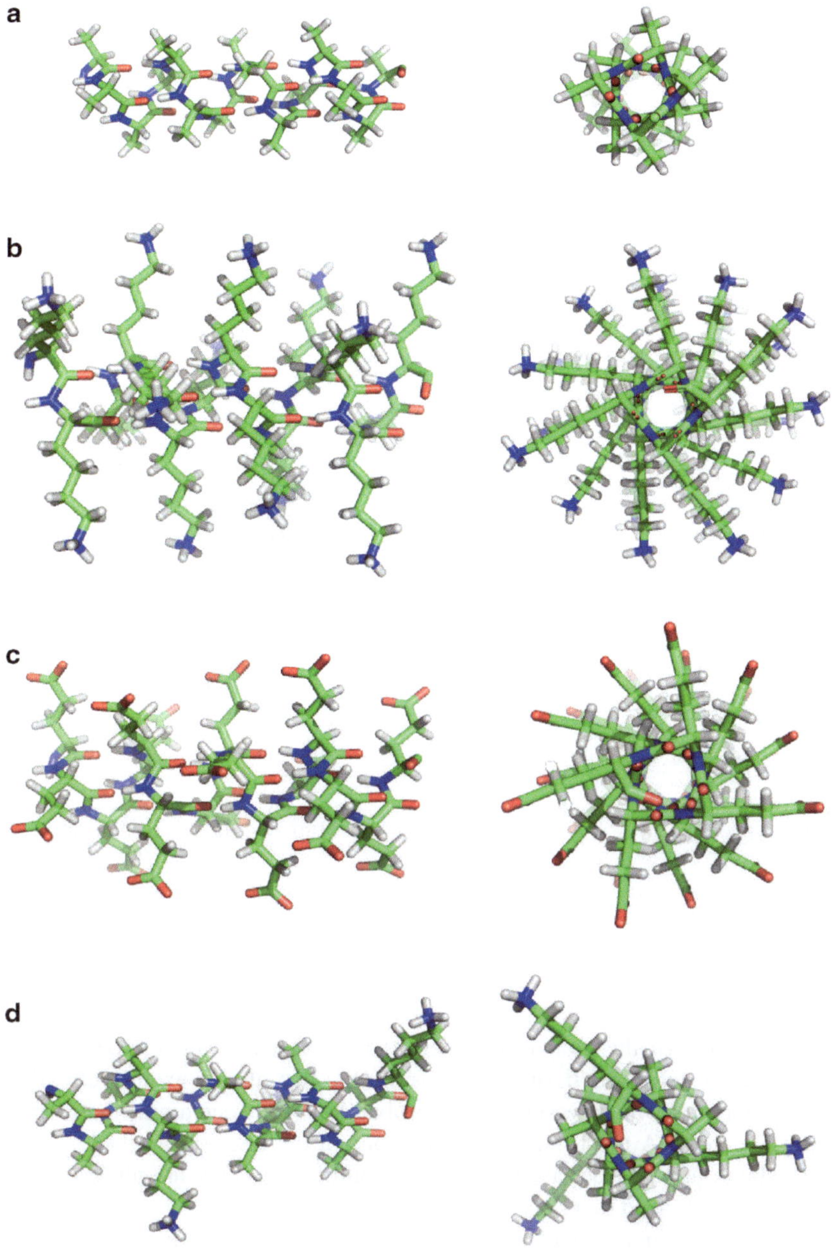

Fig. 2 Structure of α-helix. (**a**) Poly-L-alanine, (**b**) poly-L-lysine, (**c**) poly-L-glutamic acid, and (**d**) (Lys-Ala$_4$)$_3$ polypeptide, all in α-helical conformations. For each polypeptide, a side view is on the *right* and a top view is on the *left*

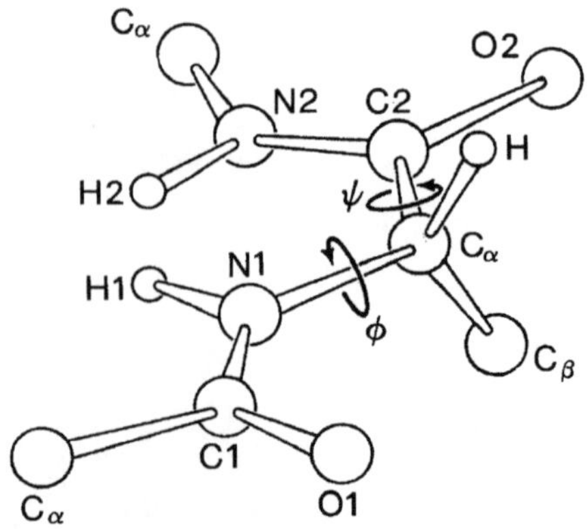

Fig. 3 Definition of dihedral angles as designated in Greek letters. Reproduced from Kyte J (2006) Structure in Protein Chemistry, Garland Science, Abingdon, UK with permission

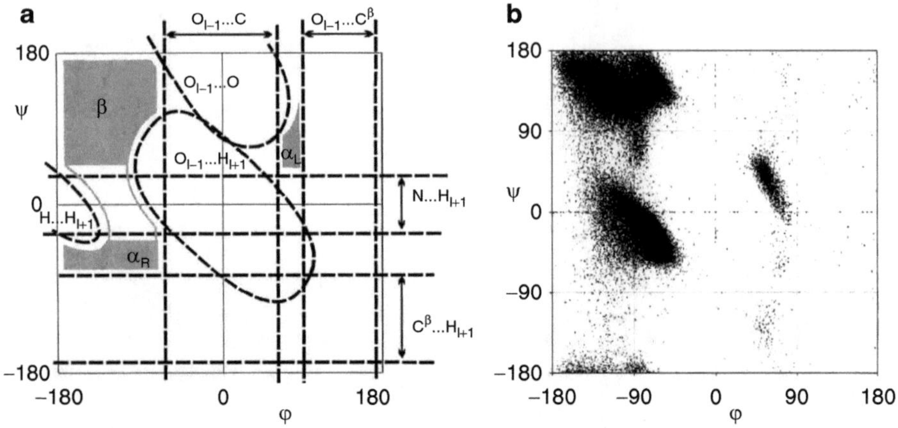

Fig. 4 Ramachandran plot. (**a**) Diagram showing regions with high energy due to steric hindrance between specified atoms on neighboring amino acid residues. (**b**) An example of dihedral angle plotting of for proteins from PDB data. (Reproduced from [37] with permission)

An inspection of the plot immediately tells us that the rather broadly defined right handed α-helix and β-sheet regions are the two almost exclusively allowed regions in the plot with lowest collision probabilities between atoms that comprise polypeptide chain. A most likely answer to one of the frequently asked questions, "Why α-helix and β-sheet all the time?," would be that they are almost the only conformations allowed to a polypeptide.

Theoretical prediction and experimental measurement of the thermodynamic stability of α-helix has been one of the central issues in biophysical research. Pioneering work was done by Zimm and Bragg [38] which was then improved by Lifson and Roig [39]. Much work has since been and is currently being done with vigorous use of statistical mechanics, molecular dynamics, and other types of computer simulation. Only a few examples are cited here [40–45]. Since the statistical mechanical treatment of helix stability has been developed primarily on a few parameters, i.e., the relative stability of the helical state against an extended state (or a randomly coiled state), often designated as s, and the degree of cooperativity, σ, estimation or determination of them are the central issue of the theoretical and experimental work on the thermodynamics of α-helix. In this review, I will only briefly touch on the equilibrium properties of α-helical polypeptides to interpret, when necessary, the data from mechanical experiments.

4 Mechanical Properties of Polypeptides and Proteins

The mechanical strength of protein molecules and protein based bio-structures has been regarded as an important concept in the study of hard and/or strong biological structures such as wool, silk, bone, etc. Work on keratin by Astbury is described above [15]. Investigations on the mechanical toughness of silk proteins including that of spiders dragline have also been actively pursued [46, 47]. Keratin is rich in α-helical conformation [48] while the secondary structures of both types of silk proteins abound in β-sheet conformation. The α-helices of keratin are bundled into a so-called "coiled coil" structure where two helices are intertwined to form a higher order helical conformation. In more general coiled coils, 2–7 (most commonly 2 or 3) [49] α-helices are coiled together like the strands of a rope through specific side chain interactions. The coiled coil form for α-keratin was proposed by Crick [50] soon after the structure of the α-helix was suggested by Pauling and coworkers [30].

Since keratin is a basic building block of mechanically strong biological structures such as nails, hair, skin, etc., interest in the mechanical properties of keratin molecules and fibers is widespread. More recently, the intracellular presence of keratin fibers as intermediate filaments in various kinds of cells also calls for the necessity to measure and understand the basic properties of keratin molecules [51].

Intermediate filament proteins are classified into subtypes I–VI of which Types I and II are, respectively, acidic and basic keratins [52]. Type III contains vimentin, the most widely distributed intermediate filament protein [53]. Desmin is another type of intermediate filament [54]. All these proteins have α-helical conformations with terminal non-helical globular domains and they are all destined to form coiled coil secondary structures. The structural and functional studies of keratin and keratin-like proteins are actively pursued in various animal models [55, 56].

In addition, a recent interest in β-amyloid like proteins with considerable thermodynamic and mechanical stability revived interest in the mechanical measurement of the physical properties of proteins [57].

Some of the β-amyloid proteins are built from a double or triple strand helix called β-helix, formed by the association of parallel β strands in a helical pattern. The structure was first found in the tail region of bacteriophages P22 and T4 where a rigid structure that would sustain the mechanical injection of DNA into bacterial cell body through cell wall and cell membrane [58, 59] and the growth property of amyloid protein was studied using TIRFM [60].

Table 1 Comparison of mechanical properties of keratinized materials and other biological and synthetic materials (modified from [10, 61] with permission.)

	Elastic modulus (GPa)	Tensile strength (MPa)	Bending strength (MPa)
Horn	6–9	140	200
Hoof	0.2–0.6	7–16	14–20
Bone	14	148	250
Abalone shell	70	170	197
Silk	5–10	300–600	
Hair	2.4	200	
Collagen	2	100	
Actin	2.5	30	
Elastin	0.002	2	
Carbon nanotube	1,300	14	
Glass	73	1,000	
Rubber	0.02	17	
Fiberglass	5.9	110	
Polycarbonate	2.4	67	

In Table 1, Young's modulus and the tensile strength (for some of them bending strength as well) are give as excerpts from [10, 61]. The values for biological materials differ considerably under dry and wet conditions. The cited values are for under more or less dry conditions.

The strength of such macroscopic materials are mainly determined by the mechanical properties of intermolecular interactions [62] and the mechanical strength of non-covalent interactions is in the range of a few hundreds of piconewtons [10]. It will be an interesting attempt to correlate the macroscopic properties of biological materials to those of constituting molecules.

5 Single Molecular Mechanical Properties of α-Helical Polypeptides

With the development of single molecule techniques using atomic force microscopy and/or optical trapping method, the experimental approach to the mechanical properties of helical polypeptides has been revived along with forced unfolding of globular proteins as stated earlier in this chapter.

Many mechanical parameters are used to characterize the strength of materials but those that can be defined and applied at the single molecular level are rather limited in number. A fundamental parameter is Young's modulus; in other words, the elastic modulus, being defined as the coefficient of the linear relationship between the stress (normal tensile or compressive force/sample cross section) vs strain (relative tensile or compressive deformation against the original sample length). In a tensile experiment at the single molecular level, the sample cross section is often ill-defined because we cannot cut samples to a desired size and the shape most appropriate for mechanical measurements. Instead, we use natural samples as they are obtained, which renders it difficult to deduce mechanical parameters as they are precisely defined. In addition, the magnitude of measurable size of the deformation is usually quite large compared with the original length of the sample. Due to such particular experimental situations, the relation between the applied force and the deformation is decidedly non-linear. Good examples are force-extension relationships in mechanical deformations of biological fibers or cell membranes.

In comparison, when the tensile force-relative extension relation is non-linear a well-defined Young's modulus cannot be obtained [14]. It is still possible to calculate a differential Young's modulus at every point of the force–extension curve and obtain the so-called tangent modulus. The tangent modulus is used to characterize the elastic behavior of soft, non-linear materials in many cases of synthetic plastics, and in most of such cases, the tangent modulus tends to become smaller as the extension of the sample becomes large and non-linear in a phenomenon sometimes called "neck formation (or necking)" in contrast to the strain stiffening mentioned earlier.

Since these material constants are defined for a perfectly isotropic and homogeneous material within an ideally elastic, small strain regime, application of such mechanical concepts to single molecules should only be done with reservations.

6 Polymer Chain Stretching

Measurement of mechanical properties of a single polymer chain, be it natural or synthetic, is a relatively new experimental and theoretical concept that requires close attention to the experimental design and theoretical interpretation [63, 64]. Although it is often true that a mechanically rigid structure is thermodynamically stable, it must be remembered that they are, in principle, different concepts. As a consequence, whereas, thermodynamic parameters such as entropy, enthalpy, and Gibbs energy are functions of states, being independent of experimental pathways from one state to another, mechanical parameters such as force and torque are not functions of states. Therefore, for example, a measured value of tensile strength, i.e., a maximum stress that a particular member can withstand before its total breakdown, depends on the loading rate of the stress, i.e., how fast the stress is applied. The numerical values of force measurements cannot therefore be directly compared to each other unless exact experimental conditions are specified.

At the molecular level, polymer chains are highly anisotropic and inhomogeneous, and their deformations under externally applied forces are not confined to the small strain regime. For example, a forced extension of an otherwise coiled chain under an externally applied tensile stress can make it reach more than ten times its original dimensions and the force–extension (analogous to stress-strain curve) is decidedly non-linear.

For such a nonlinear extension of polymer chain as stated above, the use of tangential modulus in place of Young's modulus was mentioned above but ambiguity remains in the determination of the cross-sectional area of the sample which is required to convert the applied force (N) to stress (N m^{-2}, Pa). This is especially true if the polymer chain before extension is in a tightly folded conformation as a globular protein because we have no means to estimate the cross-section of the molecule during the process of extension. We, therefore, resort to a force-extension (*F–E*) curve without normalization with respect to the sample cross section rather than a better defined stress-strain curve to characterize polymer chains from mechanical point of view. In many cases of homologous polymer chain extension, overlapping of *F–E* curves gives rise to a single master curve after normalization with respect to the full extension length because the intra-chain segmental interactions are of the same nature for all the molecules.

A good example of such a master curve is one for a group of structureless chains whose mechanical behavior is well approximated by a worm like chain (WLC) model. Experimental examples are those of polyethylene glycol [65], denatured proteins [66], and certain types of polysaccharides [67]. Some examples are given in Fig. 5.

The average mechanical behavior of the structureless polymer chain is often approximated by that of either a freely jointed chain (FJC) or a WLC model. In the former model, a large number of stiff segments usually of the same length are connected next to each other by universal joints to comprise a flexible chain and, in the latter case, a polymer chain is modeled to be continuously flexible in contrast to the freely-jointed chain model that is flexible only at the joints between discrete segments. The FJC model has the segmental length and the WLC model has the persistence length as adjustable parameters. Here we are concerned with the *F–E* curves predicted by WLC model with the persistence length as a parameter. For a polymer chain of length L_0, the path (distance along the chain) of the polymer is parameterized as s and $r(s)$ is defined as the position vector along the chain [10].

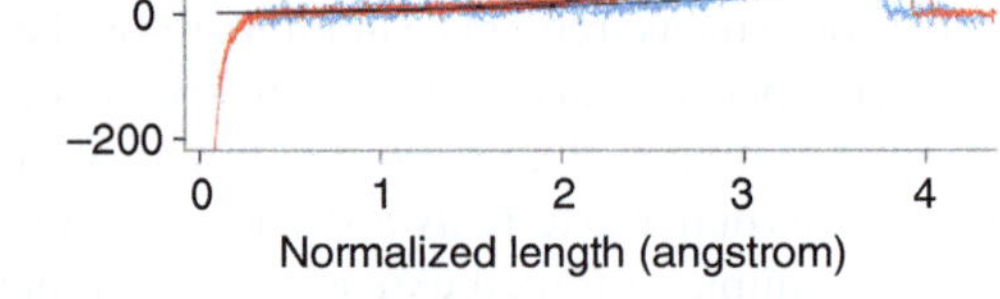

Fig. 5 *F–E* curves of PEG and denatured protein. (**a**) Polyethylene glycol (PEG) extension in PBS and organic solvent. In hexadecane, the chain is in disordered and in PBS it is in helical conformations. *Dots* are experimental results and the *solid lines* represent fitting curves of the WLC model. (**b**) *F–E* curve of denatured protein in 6 M guanidinium chloride with fitting curve of the WLC model in *gray* (originally *green*) lines (Reproduced from [65, 85] with permission)

Then the unit tangent vector along the chain, $t(s)$is defined as $\delta r(s)/\delta s$. The persistence length, p, is the characteristic length of exponentially decaying self-correlation function of $t(s)$ as presented below. By using p as the sole parameter, the force-extension relationship of the WLC is given by the following fitting curve [4]:

$$< \vec{t}(0) \cdot \vec{t}(s) > = < \cos\theta(s) > = \exp(-s/p) \tag{1}$$

By using p as the sole parameter, the force–extension relationship of the WLC is given by the following fitting curve [4]:

$$f = \frac{k_B T}{p}\left(\frac{1}{4}\frac{1}{(1-L/L_0)^2} - \frac{1}{4} + \frac{L}{L_0}\right) \tag{2}$$

where L and L_0 are, respectively, the end-to-end distance and the contour length of the chain. When there is a deformation of covalent bonds, including those of bond angle and bond length, the force–extension relation becomes

$$f = \frac{k_B T}{p}\left(\frac{1}{4}\frac{1}{(1-L/L_0+f/\kappa_0)^2} - \frac{1}{4} + \frac{L}{L_0} - \frac{f}{\kappa_0}\right) \tag{3}$$

where κ_0 is the averaged covalent stretch modulus of the polymer chain.

In this treatment of randomly coiled chain, the stiffness of a chain defined as the slope of the *F–E* curve is higher for those having shorter persistence length and vice versa. At first glance, it is against our intuition because a chain with longer persistence length is usually referred to as stiff. In the case of DNA, for example, a double stranded DNA (dsDNA) has a longer persistence length of about 50 nm than that of a single stranded DNA (ssDNA) in the range of 1.3 nm [68] and dsDNA is said to be stiffer than ssDNA but the stiffness of the chain measured as the slope of the *F–E* curves is higher for ssDNA of similar length to dsDNA (see Sect. 10).

7 Relative Stability of α-Helix

Stability of α-helix has been studied extensively both from an experimental as well as from a theoretical point of view. Work on the stability of alanine based polypeptides has been pushed forward mainly by Baldwin and his collaborators [69–71]. Poly-alanine itself is not soluble in aqueous buffers but, when hydrophilic residues such as Glu, Gln, Asp, Asn, and Lys are inserted in every fifth or so position, the resulting polypeptides are readily soluble. All these alanine based polypeptides form α-helical conformation with a moderately high propensity as determined by CD spectrum. For relatively short polypeptides, the temperature induced transition between helical and random coil states is rather a gradual one. In aqueous mixture with trifluoroethanol (TFE), these polypeptides achieve progressively higher content of α-helical conformation.

The fundamental parameters, namely σ, the helix nucleation constant, ΔH_0, the enthalpy change on helix formation, and s (0°C), the average helix propagation parameter at 0°C, were all determined by Scholtz et al. by fitting the theoretical transition curve to the experimental ones [72].

8 Mechanical Stretching of α-Helical Polypeptides

8.1 Poly-L-Lysine

Lantz et al. reported the results of forced stretching of an α-helical poly-L-lysine in an alkaline buffer solution using an AFM equipped with a magnetically oscillated cantilever for the measurement of stiffness change of the sample by detecting the change in cantilever oscillation frequency [73, 74]. The measured stiffness ranged between 0.03 and 0.12 N m^{-1}. Measurement of stiffness during the chain stretching helped to identify events corresponding to single molecule stretching from multiple stretching events. Force curves obtained showed a gradual increase in the tensile force as the polypeptide chain was elongated and overlapped with a few sharp force peaks which were interpreted as representing sudden rupture events of clusters of peptide hydrogen bonds. In Fig. 6, some of the representative force curves from their experiments are shown.

From the area under such peaks, they calculated the enthalpy of hydrogen bond rupture as 20.2 kJ mol^{-1} (approximately 5 kcal mol^{-1}) which was claimed to be in close agreement with the value reported by Dixon et al. [75]. The latter value was the result of quantum chemical calculation for an amide hydrogen bond in vacuum. These values are approximately five times larger than the experimentally estimated value of 4 kJ mol^{-1} obtained by applying the helix-coil transition theory to the temperature dependent change in the helicity of several alanine based polypeptides [72, 76]. The assignment of the sharp breaks in otherwise smooth *F–E* curves is, therefore, not clear at this moment.

8.2 Poly-L-Glutamic Acid

Idiris et al. used synthetic poly-L-glutamic acid with a C-terminal cysteine residue to study the stretching mechanics of α-helical vs random coil polypeptides [77]. Poly-L-glutamic acid is known to form a regular α-helix under acidic conditions where most of the side chain carboxy groups are protonated but turns into a randomly coiled state upon deprotonation at neutral pH, undergoing a sharp helix-coil transition between pH 4 and 5 [78]. Poly-L-glutamic acid was end grafted to a gold coated mica surface through the crosslinker, SPDP, which is reactive toward the -SH group of the terminal cysteine residue. The AFM tip was modified with the amino

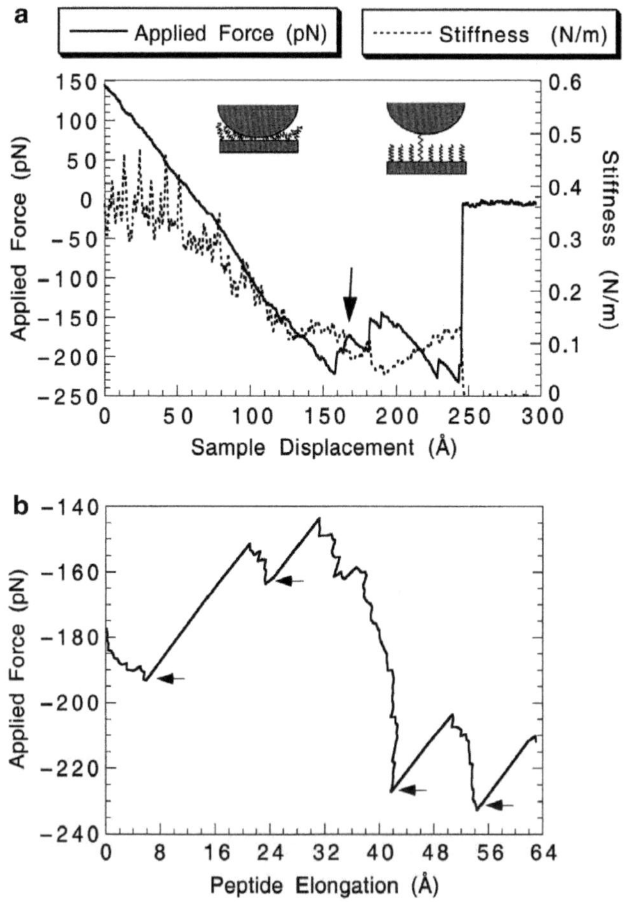

Fig. 6 Extension of poly-L-lysine. (**a**) Force curve of polypeptide extension with simultaneous recording of chain elongation and stiffness. (**b**) Force curve showing sudden relaxation of force at force peaks (Reproduced from [73] with permission)

reactive crosslinker disuccinimidyl suberimidate (DSS) and it was approached to the polypeptide on the substrate. To avoid non-specific adhesion of the sample polymer to the substrate, polyethylene glycol linker was used to keep the polymer lifted from the substrate. The *F–E* curves thus obtained were first subjected to subtraction of the polyethylene glycol extension part which was obtained in a separate experiment and the resulting *F–E* curves were normalized with respect to the maximal extension length. Representative *F–E* curves are reproduced in Fig. 7.

The *F–E* curves at different pHs showed strongly non-linear increase of the tensile force as a function of the extension of the chain. It is evident that the slope of the force curve increased as the pH was lowered, i.e., as the helical content was increased. The *F–E* curves at higher pH were very similar to the one given by the WLC model except in the middle region of the curve where the poly-L-glutamic

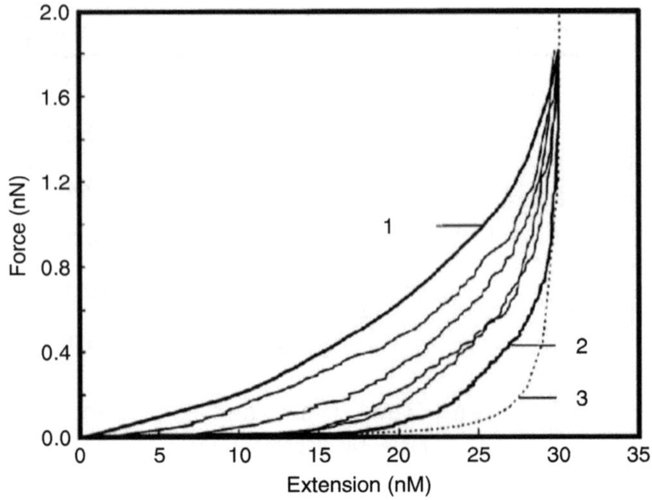

Fig. 7 *F–E* curves of poly-L-glutamic acid at different pHs. Curve 1: pH 3.0, Curve 2: pH 8.0, and Curve 3: fitting curve of WLC model. Other curves are *F–E* curves obtained at different pHs between 8.0 and 3.0 (Reproduced from [77] with permission)

acid curve is slightly higher than the former. The approximate stiffness of the chain at the middle of stretching can be calculated to be 0.04 N m^{-1} for an almost fully helical chain at pH 3.0, whereas the stiffness of a disordered chain at pH 8.0 is less than 0.003 N m^{-1}. Its Young's modulus in the early part of chain extension at pH 3.0 was estimated to be about 3 GPa assuming that the radius of α-helix main chain is 0.2 nm. If we take 0.25 nm as radius, the modulus becomes about 2 GPa. In [19], Kojima et al. reported the value of 10 GPa as the Young's modulus of tropomyosin, a coiled coil protein involved in muscle contraction. Considering that in tropomyosin structure, two helices are strengthened through side chain interactions, 3 GPa for an isolated single α-helix is within reasonable agreement. One important difference is in the value for the radius of α-helix. Idiris et al., considering only the main chain contribution, assumed it to be about 0.2 nm, whereas Kojima et al. used the value of 0.5–0.6 nm. The latter value was probably the radius including peptide side chains. The stiffness of the chain continued to increase up to the rupture point. Since the *F–E* curves rarely showed breaks as were observed for poly-L-lysine, most of the stretching work done to the chain was stored as a strain energy. The non-linear elastic energy stored in the chain at the end of stretching was reported to be about 5,000 kJ mol^{-1} (50–60 kJ/residue/mol) at pH 3.0 and 1,000 kJ mol^{-1} (10–12 kJ/residue/mol) at pH 8.0.

The results in Fig. 7 confirmed that the helical poly-L-glutamic acid is significantly stiffer than randomly coiled chain at all range of extension. In both cases, the *F-E* curves were finally ruptured at a high force of >1.8 nN due to the breakdown of the crosslinking system.

The reversibility of the curve was confirmed by repeating the approach and retraction cycle for more than 60 times without breaking the crosslinker. Although decidedly non-linear, this is almost an ideal coil spring behavior at the molecular level.

8.3 Calmodulin and Spectrin as α-Helical Proteins

Calmodulin is a small α-helical protein with two globular subdomains connected by a single α-helical rod as shown in Fig. 8.

We pulled this protein from its two ends and obtained *F–E* curves for two types of calmodulin, one with (holo) and the other without (apo) Ca^{++} ions [27]. The *F–E* curves shown in Fig. 9 clearly show differences in the mechanical properties between holo and apo calmodulin, the former having a sharp force peak after extension of about 70% of the chain length.

The *F–E* curve of apo calmodulin showed a lower stiffness compared with that of holo-calmodulin, but it was still clearly higher than that of the WLC model up to approximately 80% extension. It was confirmed by CD spectroscopy that the modified calmodulin had the same level of helicity as the native protein but not longer oligomers. Dimer of calmodulin gave similar *F–E* curves with a force peak

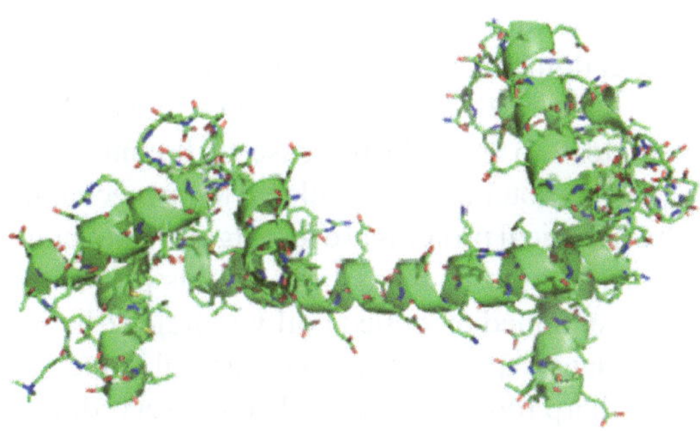

Fig. 8 Structure of α-helical calmodulin. Two globular subdomains at the C- and N-termini are connected by a single α-helical chain

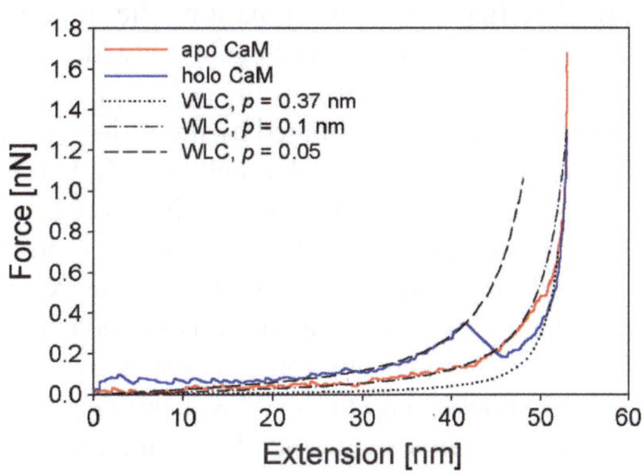

Fig. 9 *F–E* curves of holo and apo calmodulin (Reproduced from [26] with permission)

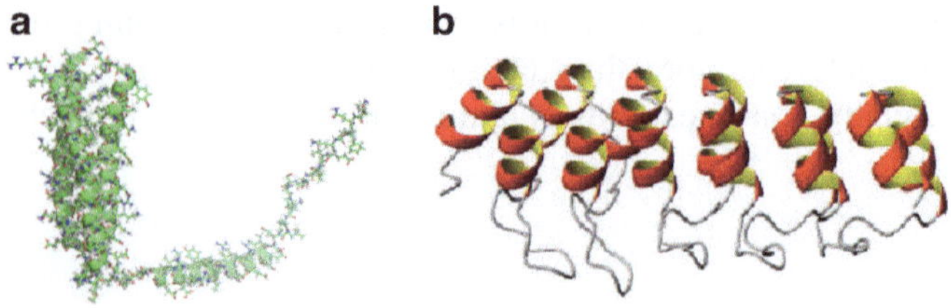

Fig. 10 Structure of (**a**) spectrin and (**b**) ankyrin repeats. Short α-helical chains are connected by linking peptides with non-regular conformations

but not those of higher oligomers. Spectrins belong to a widely distributed protein family and, among them, one best known member is in red blood cells as the major component of the cytoskeletal structure [13]. It is a highly α-helical protein with three helical rods assembled into a globular tertiary structure and this unit structure is repeated several times along the major axis of the molecule (Fig. 10).

When this protein was constructed in a series just like it was in the titin model introduced above, a series of force peaks were recorded again in a similar manner as in the case of titin and other tandemly constructed proteins as reproduced in Fig. 11 [79, 80].

In this and other cases of stretching experiments of tandemly repeated protein of α-helical conformation, the peak force values were noticed to be clearly less than in the case of β-sheet proteins such as titin and immunoglobulin repeats. Not only β-sheet domains in tandemly repeating proteins such as titin but also intramolecular cores made of β-sheets were shown to have high values of tensile strength [25, 27], even close to 1 nN in the case of bovine carbonic anhydrase II [25]. In general, α-helical proteins are considered to be mechanically less rigid than β-sheet proteins. Why α-helical proteins are less rigid is not yet fully explained. A study on the helical regions of myosin sheds light on this problem to some extent by comparing force curves obtained on coiled coil vs single α-helical chain variants [81, 82].

In many reports on the mechanical stretching of coiled coil type α-helical proteins, sharp force peaks have been observed. But in such cases, the force peaks have been attributed to fracture events of side chain interactions that stabilize α-helices through tertiary structure formation. The situation is similar in the breakdown of the β-sheet structure in titin molecules as widely studied by atomic force microscopy. The conformational states of the polypeptides after the breakdown of tertiary structures have been approximated by the fitting curve of the WLC model assuming that little secondary structural elements are left. In the case of β-sheet proteins, it is probably correct, but for α-helical proteins there is a good chance that helical secondary structure is retained after the breakdown of the tertiary structure. Unfolding of bacteriorhodopsin under an applied force is an exceptional case in this respect [83, 84]. Since the body of the protein is constrained in the lipid bilayer membrane, stretching of α-helical segments by force is expected to pull out the peptide chain sequentially from the attached end to the AFM cantilever unraveling the helical conformation at the interface between the membrane and the solvent.

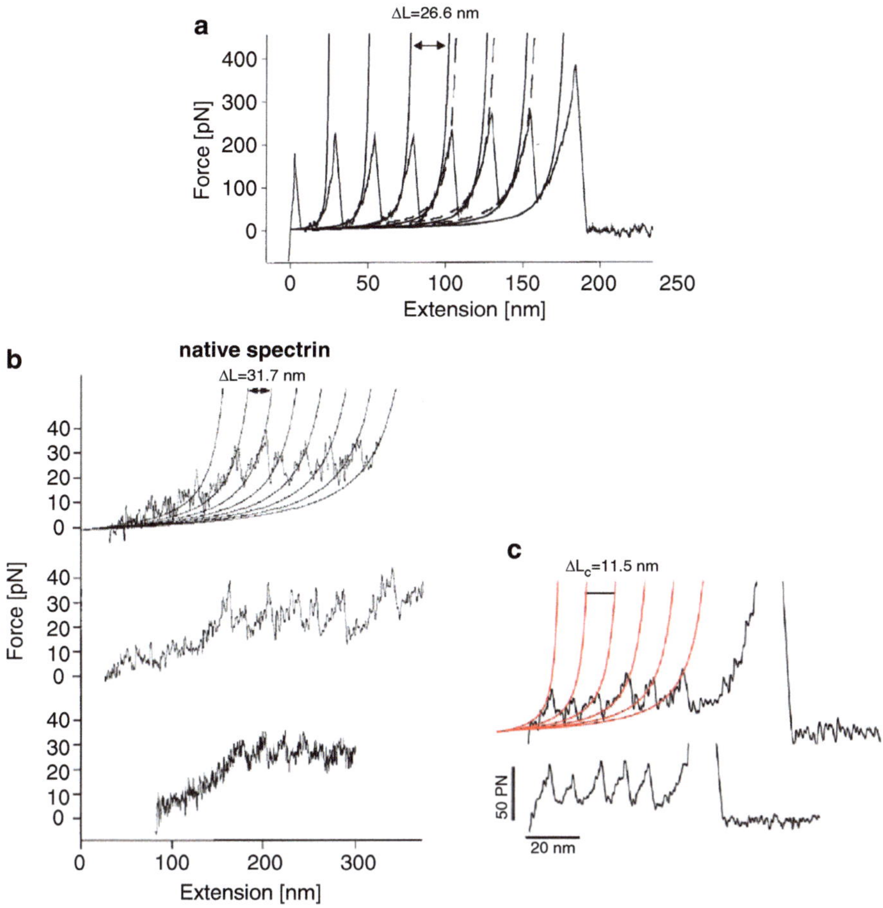

Fig. 11 *F–E* curves of titin (**a**), spectrin repeat (**b**) and ankyrin (**c**) repeats proteins (Reproduced from [79] and [80] with permissions)

*8.4 Poly-*L*-Alanine*

Afrin et al. investigated the mechanical properties of a polypeptide having the primary sequence of CK(AAAAK)$_{10}$C (to be abbreviated as CKA where C, K, and A represent amino acid residues, cysteine, lysine, and alanine, respectively) in phosphate buffered saline (PBS) and in 50% TFE [85]. The CKA polymer chains were grafted on to a silicon surface after activating it with amino silanizing and crosslinking reagents. The CKA chains were spontaneously oligomerized during storage due to disulfide bridge formation between the terminal cysteine residues, a convenient result for force measurement, because the chain length of the original polymer was not long enough to guarantee reliable extension experiment. Pulling experiments were performed on CKA oligomers and the resulting tensile force against the relative extension data were plotted for the chain as reproduced in Fig. 12. The *F–E*

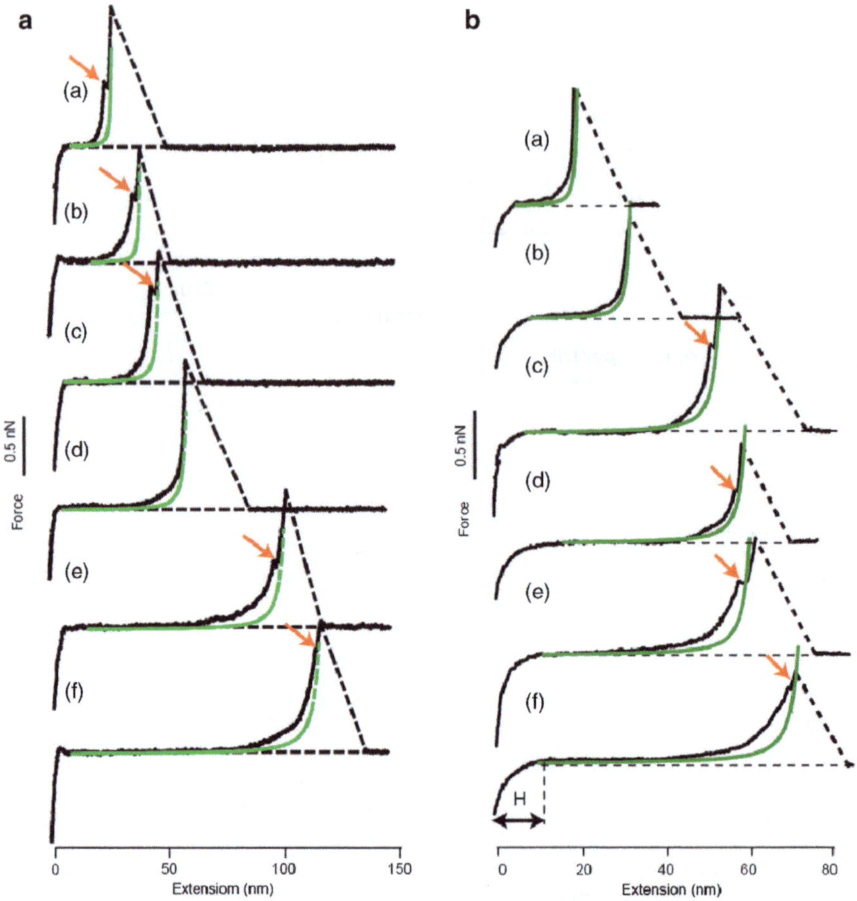

Fig. 12 *F–E* curves (*black*) of CKA in neutral buffer, (**a**) in PBS, (**b**) in PBS + 400 mM NaCl. *Gray* (originally *green*) lines are fitting curves of the WLC model and *arrows* point to the position of small inflection points (Reproduced from [85] with permission)

curves in Fig. 12 were obtained in aqueous PBS buffer at pH 7.4 with and without a high concentration of NaCl. Those in Fig. 13 were obtained in 50% TFE kept at pH 7.4 with PBS.

The results of CD spectroscopy showed that the helicity was 60–70% in the former solvents and close to 80% in the latter solvent. In both cases, the tensile force was low (lower than 20 pN which was the amplitude of thermal noise) until a steep increase was observed as the chain length limit was approached. In the case of polypeptide with 80% helicity, the *F–E* curves are of a higher range than those in aqueous buffer or that of WLC but the α-helix of CKA peptide is still a labile structure with a low resistance against tensile force.

The convex curve in the initial part of the extension (on the left edge of the curves) was due to the repulsive force exerted from the compressed polymer against

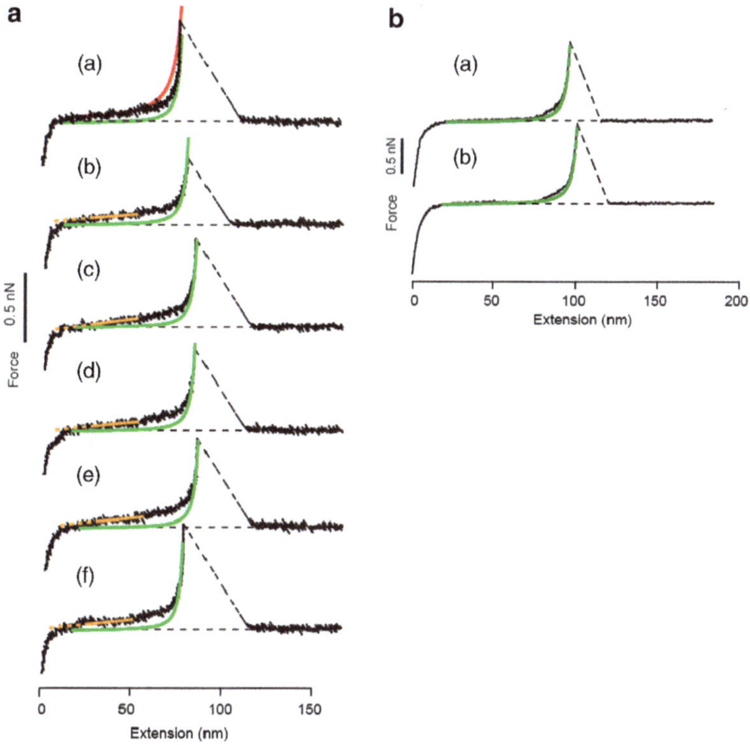

Fig. 13 *F–E* curves of CKA, (**a**) in 50% TFE, (**b**) in 6 M guanidinium chloride (Reproduced from [85] with permission)

the AFM probe. It is clearly shown that the polymer chain formed a brush region of 10–20 nm in thickness above the silicon substrate, indicating little adsorption of CKA to the substrate. No PEG crosslinkers were used in all cases. The *F–E* curves in 50% TFE at pH 7.4 given in Fig. 13 show small but clear deviation from that of the WLC model, especially in the middle region of relative extension.

The low value of tensile force of CKA compared with those of poly-L-glutamic acid or poly-L-lysine was enigmatic at first. When they were compared with those of forced extension of calmodulin, it appeared to be due to partially helical polypeptide made of amino acid residues of small or moderate sized side chains.

To facilitate understanding the low tensile force behavior of CKA and calmodulin upon forced uniaxial extension from their two ends, a steered molecular dynamics simulation (SMD) was carried out by the same group. The SMD tries to simulate the uniaxial pulling experiment by inserting a virtual harmonic spring between one end of the sample polymer and a hard surface representing cantilever or substrate surfaces.

In Fig. 14, some results of steered molecular dynamics simulation are given.

The *F–E* curves of CKA at three different pulling speeds, i.e., 0.1 nm ps^{-1}, 0.01 nm ps^{-1}, and 0.001 nm ps^{-1}, became more flattened as the pulling speed was

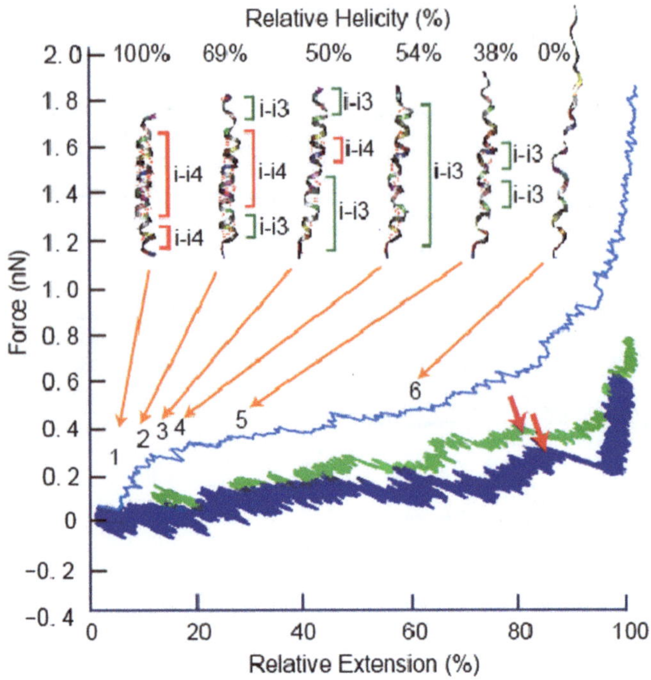

Fig. 14 Result of SMD simulation on CKA peptide stretching. Three curves correspond, from *top* to *bottom*, to the pulling speed of 0.1 nm ps^{-1}, 0.01 nm ps^{-1}, and 0.001 nm ps^{-1}. The snapshots on the *top* shows a series of schematic unfolding events at specified point on the *top curve*. *Arrows* point to the positions where force relaxations were observed (Reproduced from [85] with permission)

decreased and if one can extrapolate the pulling speed to the range commonly used in experiments, i.e., approximately, 100 nm s^{-1} to 1 μm s^{-1}, a very low tensile force would be expected. The trajectory of each peptide group on a Ramachandran plot was traced to clarify the pathway of α-helix to an extended conformation (Fig. 15).

It is clear that the majority of the peptide groups made their way to β-sheet by overcoming the energy barrier at around $\phi = -160°$ and $\psi = 0°$ region, i.e., cis configuration with respect to ψ angle. Although cis configuration is energetically unfavorable route, it is still the lowest energy barrier on the pathway from α-helix to β-sheet conformation.

9 Different Scenarios of Hydrogen Bond Breakdown

By computer simulation, it has been predicted that at the beginning of polypeptide stretching under tensile force, $i \rightarrow i+4$ main chain hydrogen bonds are replaced by $i \rightarrow i+3$ hydrogen bonds corresponding the helical geometry of 3_{10} helix [42, 85].

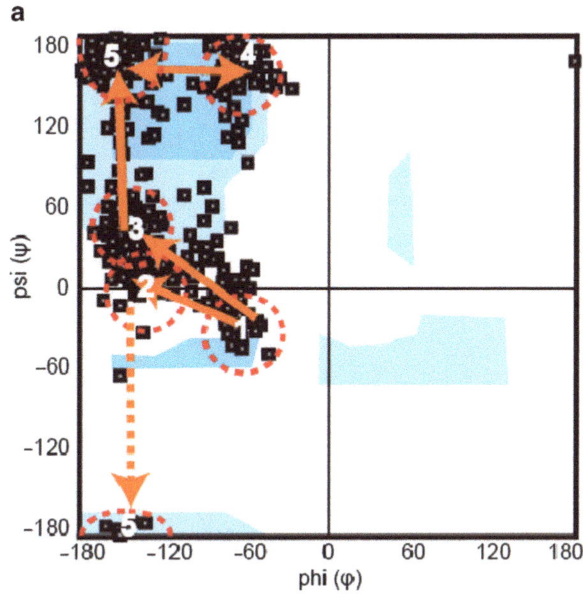

Fig. 15 Ramachandran plot of the sequential change of dihedral angles during the forced stretching of CKA polypeptide from position 1 to positions 4 and 5 through 2 and 3 (Reproduced from [85] with permission)

The latter helix has a per residue displacement of 0.2 nm and three residues per turn, elongating the original α-helix by 33% in length. Compared with the final extension to β-sheet by 160–170% from the original α-helix, a considerable further extension is needed. We, therefore, have to bring about rupture events of peptide hydrogen bonds in order to stretch a helical polypeptide from there on, but it is not clear whether such an event would be observed in an *F*–*E* curve as a sharp force peak suggesting a brittle fracture, or as a gradual change in the slope of the curve characterized by, for example, the appearance of a plateau region.

Rohs et al. presented two alternative pathways for α-helix unraveling, uniform and turn-by-turn unraveling as reproduced in Fig. 16 [42].

In the uniform stretching mode, change in energy is smooth but the force curve is characterized by a pronounced plateau and, in the turn-by-turn mode, the energy change has many breaks but the force curve shows a monotonous increase upon stretching. At least from the result given in the literature, no sudden relaxation of force was indicated.

The hydrogen bond is undoubtedly the most important element in the formation, geometrical characterization, and stabilization of α-helix but neither its mechanical strength nor the breakdown pattern due to tensile stretching have been experimentally examined in detail. According to [72], the enthalpy of main chain hydrogen bond formation is approximately in the same range for poly-L-glutamic acid, poly-L-lysine, and alanine based polypeptides, being approximately 1–1.5 kcal $mole^{-1}$ (4–6 kJ mol^{-1}) with very little effect from side chain differences. In our mechanical unfolding experiments, however, large differences were found in the force required to stretch poly-L-lysine, poly-L-glutamic acid, and CKA, suggesting that the observed differences are most likely reflect the difference in the side chain properties.

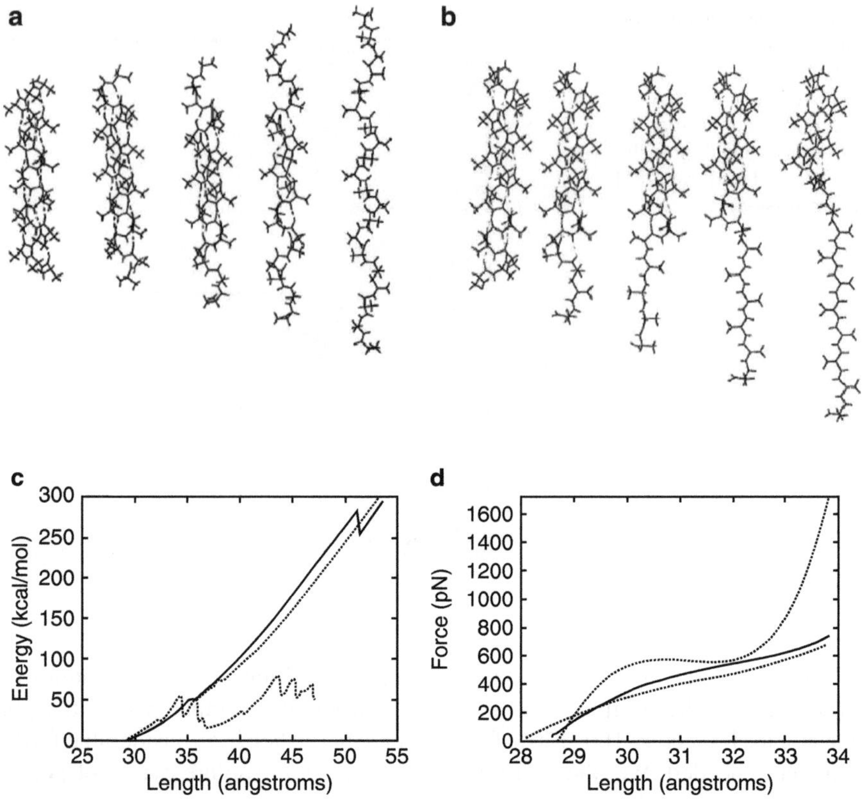

Fig. 16 Two different modes of α-helix unfolding. (**a**) Uniform and (**b**) turn by turn unfolding. (**c**) Energy and (**d**) force change during stretching (*solid line*: uniform stretching of an $(Ala)_{20}$ α-helix; *dotted line*: uniform stretching of an $(Ala)_9$-Pro-$(Ala)_{10}$ α-helix; *dashed line*: turn-by-turn stretching of $(Ala)_{20}$ α-helix) (Reproduced from [42] with permission)

The tensile force should also depend on the speed of stretching polypeptides. In the case of helix unraveling by force, several different pathways of hydrogen bond breakdown can be found in the literature. One possibility is that, during unraveling of the helix, the peptide hydrogen bonds in one or a small number of helical parts are sequentially stretched and broken at the interface with the unfolded parts. This scenario assumes a slow interchange between helical and random coil parts in equilibrium. In this case, hydrogen bonds are broken down at a constant rate, giving rise to an *F–E* curve with a plateau force in its middle section.

The other extreme case is one of rapid interchange between helical and coiled parts where formation and breakdown of peptide hydrogen bonds proceed rapidly compared with the time scale of chain stretching due to the applied tensile force [86, 87]. Stretching of the polypeptide chain, in this case, takes place preferentially from the instantaneously formed random coil parts because of the lower stiffness of such parts compared with the helical parts. In Fig. 17, a theoretical curve for polypeptide extension under tensile force is given.

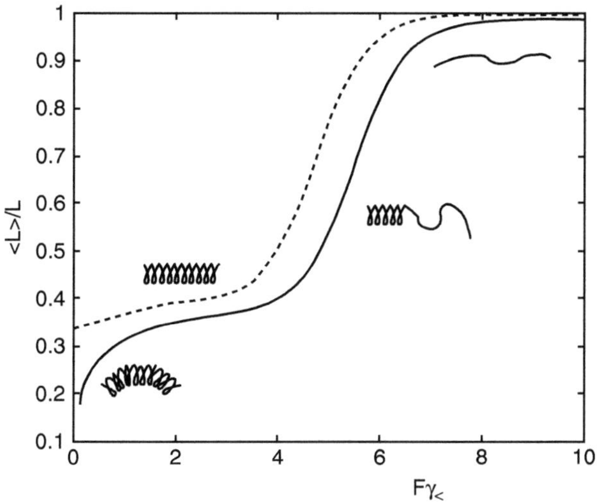

Fig. 17 Theoretical *F–E* curves incorporating a rapid helix-coil fluctuation during chain stretching. In this figure the ordinate is extension and the abscissa corresponds to tensile force. Symbols are: $< L >$, average end to end distance; L_0, the total contour length; F, tensile force; $\gamma_<$, the relative bending modulus of the chain (Reproduced from [86] with permission)

The potential curves of various types of hydrogen bonds have been studied and, in the case of the peptide hydrogen bond, the equilibrium length between the hydrogen and the oxygen in >NH–O=C< is reported to be approximately 0.195 nm but the anharmonic potential based on dipole-dipole interaction is rather wide and the precise breakdown point is difficult to predict [88, 89].

Another interesting theoretical prediction is that a partially helical chain, i.e., a polypeptide chain near the bulk helix-coil transition, should increase its helicity upon tensile stretching by the application of external force due to the loss of conformational entropy of coil state under uniaxial tension [90–92]. This interesting prediction has been substantiated in a macroscopic experiment using a block of collagen gel and an optical activity measurement system [93, 94] but not yet verified at the single molecular level mainly due to the lack of a method capable of measuring the helical content of a single polypeptide chain. Whether it can be verified in the near future depends on the development of such a method as an extension of, for example, today's single molecule Raman spectroscopy. Otherwise, it is difficult to associate changes in *F–E* curves obtained from mechanical measurement to a change in helical content of a single chain. Whether the helix that is expected to be formed under a tensile strain would be α-helix or one of more elongated helices such as 3_{10} helix is an interesting question. An increase in the helical content in a chain should make the chain stiffer but the chain may become either shorter or longer depending upon the number of residues in a chain as well as on the extent of non-ideal segmental interactions either attractive or repulsive on average.

There are at least two factors to be considered when we try to understand the conformational states of the sample polypeptide during its forced stretching. First, the experimentally determined rate constants for helix formation and breakdown are quite large, both at least larger than $10^3\,s^{-1}$ and can be as large as $10^{7-8}\,s^{-1}$ [95–100]. The rate of pulling of helical polypeptides by AFM is usually slower than the time constants for the fluctuation of the chain conformation between helix and coil as discussed below. When the pulling speed of polypeptide chain under AFM experiments is $1{,}000\,nm\,s^{-1}$, it takes 1 ms to stretch the chain by 1 nm (average of $0.01\,nm\,s^{-1}$ per residue for a polypeptide with 100 residues) and 0.1 ms to stretch 0.1 nm ($0.001\,nm\,s^{-1}$/residue). If a few tenths of a nanometer stretching takes place at a single helix-coil interface along the chain due to a spontaneous stress concentration there, the rate tends to overlap the slower range of the rate of fluctuation but such a concentration of stress is probably highly unlikely. Thus, we assume that stretching of a polypeptide is forced on a thermodynamically fluctuating chain around its equilibrium between helical and coil states with constantly changing interfaces between the two conformations.

10 Comparison with DNA Stretching

Stretching experiments of a single polymer chain was, in fact, initiated by Bustamante and his colleagues on dsDNA using the force measuring magnetic beads or optical trap method [3, 5]. In the first report using magnetic beads, *F–E* curves were obtained for individual DNA molecules at three different salt concentrations with forces between 10^{-14} and 10^{-11} newtons. Deviations from the force curves predicted by the FJC model suggested that DNA had significant local curvature in solution. In the second experiment using an optical trap method, a single molecule of double-stranded DNA (dsDNA) were stretched close to its contour length as dsDNA under a longitudinal stress of approximately 65 pN. As the force was increased above this value, the dsDNA underwent a highly cooperative transition into a stable form with 0.58 nm rise per base pair compared with that of 0.34 nm for the B form, that is, 70% longer than B form dsDNA. During the transformation of dsDNA to the overstretched DNA, the tensile force remained constant giving rise to a long plateau in the *F–E* curve before the chain reached the maximum contour length for the overstretched state. This new state of dsDNA was called an "overstretched DNA." In their experiment, fitting of the *F–E* curves of individual molecules of single-stranded DNA gave a persistence length of 0.75 nm and a stretch modulus of 800 pN. Thus, from the point of view of the present review, dsDNA was shown to provide a constant-force spring in addition to a normal WLC spring with the entropic origin of tensile resistance. Compression mechanics of dsDNA has been studied, for example, from the interest of its confinement in a small space in the biological application of MEMS technology [101, 102]. Tegenfeldt et al. showed that genomic-length DNA molecules imaged in nanochannels have an extension along the channel that scales linearly with the contour length of the polymer, in agreement with the scaling arguments developed by de Gennes for self-avoiding confined polymers [103, 104]. This fundamental relationship allowed them to measure the contour length of a single

DNA molecule confined in the channels. Reccius et al. reported the time constant for the relaxation of electrophoretically compressed DNA to a more extended form in a narrow channel when the electric field was turned off [102]. The channel provided a space for confinement of DNA by the presence of a narrow constriction near the positive electrode. The authors developed the following equation relating the experimentally measurable relaxation time, τ_d, to the frictional constant of DNA in the channel, ξ, and the persistence length, b:

$$\tau_d = \frac{2}{9}\beta a^3 n\xi N^2 = \frac{4}{9}\beta\xi b L^2 \tag{4}$$

where $\beta = (k_{\mathrm{B}}T)^{-1}$ and L is the contour length. By using an optical method to record the expansion of compressed dye-conjugated DNA from a narrow constriction, they measured $\tau_d = 66.7\mathrm{s}$ and, with a reasonable estimate on the frictional coefficient, they obtained the persistence length $b = 28 \pm 3\mathrm{nm}$ for the monomer of λ DNA, approximately a half of what was expected for a dye-conjugated dsDNA. The reason for the shorter b value was explained by forced bending of DNA strand and the possible formation of kink structures in the compressed state.

11 Scale Effect at Nanometer Scale Springs

When looked at as a ordinary coil spring, the α-helix is one of the smallest and it will be interesting to see how the well known formula to estimate the spring constant (stiffness) of a coil spring can be applied. To start with, we realize that the spring constants (stiffness) of cantilevers, beam springs and coil springs are not scale free.

Suppose a coil spring is made of a material with a shear modulus of G, and the spring has the effective number of turns, N, radius of the coil, R, and radius of the wire, r, then the spring constant or stiffness, k, is given as follows [105, 106]:

$$k = \frac{1}{4N}\frac{Gr^4}{R^3}\left[1 + \frac{1}{2}\left(\frac{r}{R}\right)^2\right] \tag{5}$$

The stiffness of the spring therefore depends on the geometrical size of the spring even if the shear modulus, the ratio of R and r, and the number of turns are kept constant. The second term in the parenthesis of the right hand side of (5) can be neglected when $(r/R)^2$ is small. To illustrate this situation, let us calculate the stiffness of two springs with the same values for N and G but with different R and r by a factor of 10^7 while keeping the ratio, R/r constant. In the case of α-helix, the coil diameter is close to 6 nm without side chains, and that of the polymer chain is in the range of 0.13–0.3 nm [107]. We take $R = 3 \times 10^{-10}\,\mathrm{m}$ and $r = 1 \times 10^{-10}\,\mathrm{m}$ in the following example in Table 2. The spring constant of the smaller spring, though it has the same geometrical proportionality with the larger one, is smaller than that of the larger one by a factor of 10^{-7}.

Table 2 Size dependence of spring constant of helical springs

	$k\,(\mathrm{N\,m^{-1}})$	N	$G\,(\mathrm{GPa})$	$R\,(\mathrm{m})$	$r\,(\mathrm{m})$
3 mm Coil spring	1.8×10^{2}	50	1	3×10^{-3}	1×10^{-3}
α-Helix	1.8×10^{-5}	50	1	3×10^{-10}	1×10^{-10}

The observed scale effect is not a special case for a coil spring, but also true for a cantilever spring whose stiffness is expressed as

$$k = \frac{3EI}{L^3} = \frac{1}{4}\frac{Ewt^3}{L^3} \tag{6}$$

where E, I, w, t, and L are the Young's modulus, second moment of the cross-sectional area, width, thickness, and length of the cantilever.

The example given above immediately tells us that the experimentally observed stiffness of the α-helix in the order of $1\text{–}50\,\mathrm{pN\,nm^{-1}}$ is quite a high value that requires a material of shear modulus of 50–2,500 GPa (130–6,500 GPa, in terms of Young's modulus). The question here arises how to reconcile the requested high shear modulus for normally soft materials like proteins whose Young's modulus is in the range of 0.1–10 GPa. Apparently the stiffness of the molecular level coil spring does not directly come from the mechanical elasticity of the material that it is composed of but comes from some other source, such as hydrogen bonds between turns, side chain interactions, or possibly from the dihedral constraints during the transition from right handed α-helix to the extended sheet structure. Probably all of these factors may contribute to endow the real α-helix its stiffness unusual for its small size, hydrogen bonds in the small deformation regime, side chain interaction and dihedral angle constraints in the medium to large deformation regimes. We would like to address these problems more precisely in the future to bridge macroscopic material science to nanometer sized material science. In the case demonstrated above, the ratio, $r/R = 1/3$ and is not small enough to guarantee the used of the equation for accurate prediction of the spring stiffness but it is still usable within a small correction. Moreover, in a large deformation regime, (5) may not be valid.

If we look at this problem as a torsional stress of a cylindrical bar, the relation between the rotational angle

$$G = \frac{TL}{\theta I_p} \tag{7}$$

where $T = FR, L, I_p = (1/2\pi)r^4$, and θ are the torque length of the shaft (bond length), polar moment of inertia, and the twist angle of the bond. If the T = pulling force $\times$ helix radius, L = bond length, r = radius of the bond, and the rotation per single bond angle is in the order of 1 radian, then is G is calculated to be in the order of 3 GPa:

$$G = \frac{50\times10^{-12}\times3\times10^{-10}\times1.5\times10^{-10}}{1\times(1.57\times10^{-10})^4} = 3\times10^{9} \tag{8}$$

The value of G thus obtained is an order of magnitude less than the value calculated above. Whether this difference is due to the inappropriateness of applying equations derived for coil springs made of isotropic and homogeneous materials or to the bad choices of some of the structural parameters is to be verified in the future.

12 Conclusion and Future Prospects

An emerging mechanical profile of the α-helical conformation of polypeptide in large deformation regime resembles those of conventional flexible polymer chains. In a small deformation regime, however, shorter α-helices of higher rigidity, either isolated or included in longer protein chains, have been a subject of theoretical studies [108, 109]. A careful study of α-helical polypeptides by NMR suggested that unfolding of α-helical amino acid residues is hardly possible in the middle of stiff long chains [110]. It is most likely that, under a large deformation scheme, the side chain-main chain and side chain-side chain interactions may affect the mechanical properties of the α-helical peptides as a source of additional rigidity. The mechanical study of α-helical peptides is still very much in its infancy from the experimental point of view and needs considerable effort in terms of technological advancement of single molecule manipulation to verify the validity of experimentally obtained results, i.e., F–E curves and chain stiffness data. Experimentally there seems to be a large dependence of the chain stiffness on side chain properties but there is a paucity of theoretical or computational support for this observation so far. A close collaboration of computer simulation with experimental work is vitally needed for the future development of the mechanics of the conformations of polypeptide chains based on various aspects of the deformation and rupture of a hydrogen bond and hydrogen bond clusters under tensile constraints.

Finally, in this connection, it may be interesting to note that the main chain-side chain interactions play a key-role in the helix-sense reversal of α-helical polyaspartic acid esters observed in conventional helicoidal solution [111]. The transition may be induced by either temperature or solvent. The primary cause responsible for such a highly cooperative transformation of α-helical hydrogen-bonded network was found to reside in a small free-energy difference in the conformational states of the side chain flanking the spiral backbone of the two opposite handedness. Electrostatic side-chain interactions in polypeptides have been proposed as a stiffening factor of helices in polypeptides having sequences of $[Glu_4Lys_4]_n$ [112, 113]. I thank Professor Robert Baldwin of Stanford University for bringing refs. [112, 113] to my attention.

Acknowledgements I thank Dr. Akihiro Abe, Professor at Tokyo Polytechnic University, for his encouragement to write this review. I would also like to thank him and anonymous reviewers for their expert comments to improve the original manuscript. I am grateful to Dr. Rehana Afrin, Associate Professor at Tokyo Institute of Technology, for her invaluable advice in the process of writing this chapter. This work was supported by a Grant-in-Aid for Creative Scientific Research (#19GS0418) from the Japan Society for the Promotion of Science (JSPS) to AI.

References

1. Breslow R (2009) J Biol Chem 284:1337–1342
2. Bhushan B (2009) Philos Transact A Math Phys Eng Sci 367:1445–1486
3. Smith SB, Finzi L, Bustamante C (1992) Science 258:1122–1126
4. Bustamante C, Marko JF, Siggia ED, Smith S (1994) Science 265:1599–1600
5. Smith SB, Cui Y, Bustamante C (1996) Science 271:795–799
6. Chu S (2003) Philos Transact A Math Phys Eng Sci 361:689–698
7. Brochard-Wyart F, Tanaka T, Borghi N, de Gennes PG (2005) Langmuir 21:4144–4148
8. Higashi N, Koga T (2008) Adv Polymer Sci 219:27–68
9. Boal D (2002) Mechanics of the cell. Cambridge University Press, Cambridge, UK
10. Howard J (2001) Mechanics of motor proteins and the cytoskeleton. Sinauer Associates Inc., Sanderland, MA
11. Ikai A (2008) The world of nanobiomechanics. Elsevier, Amsterdam, pp 23–41
12. Wang Y, Discher D (2007) Cell mechanics (methods in cell biology 83). Academic Press, San Diego, CA
13. Mofrad MRK, Kamm R (eds) (2006) Cytoskeletal mechanics: models and measurements. Cambridge University Press, Cambridge, UK
14. Fung Y (1993) Biomechanics: motion, flow, stress, and growth. Springer, New York
15. Astbury WT, Dickinson S (1935) Biochem J 29:2351–2360
16. Ethier C, Simmons C (2007) Introductory biomechanics. Cambridge University Press, Cambridge, UK
17. Kinney JH, Marshall SJ, Marshall GW (2003) Crit Rev Oral Biol Med 14:13–29
18. Hearle J (2007) J Mater Sci 42:8010–8019
19. Kojima H, Ishijima A, Yanagida T (1994) Proc Natl Acad Sci U S A 91:12962–12966
20. Deguchi S, Ohashi T, Sato M (2006) J Biomech 39:2603–2610
21. Suresh S (2003) Fatigue of materials, 2nd edn. Cambridge University Press, Cambridge, UK
22. Florin EL, Moy VT, Gaub HE (1994) Science 264:415–417
23. Mitsui K, Hara M, Ikai A (1996) FEBS Lett 385:29–33
24. Rief M, Gautel M, Oesterhelt F, Fernandez JM, Gaub HE (1997) Science 276:1109–1112
25. Alam MT, Yamada T, Carlsson U, Ikai A (2002) FEBS Lett 519:35–40
26. Hertadi R, Ikai A (2002) Protein Sci 11:1532–1538
27. Hertadi R, Gruswitz F, Silver L, Koide A, Koide S, Arakawa H, Ikai A (2003) J Mol Biol 333:993–1002
28. Clausen-Schaumann H, Seitz M, Krautbauer R, Gaub HE (2000) Curr Opin Chem Biol 4:524–530
29. Butt H, Cappella B, Kappl M (2005) Surf Sci Rep 59:1–152
30. Pauling L, Corey RB (1951) Proc Natl Acad Sci U S A 37:729–740
31. Kendrew JC, Bodo G, Dintzis HM, Parrish RG, Wyckoff H, Phillips DC (1958) Nature 181:662–666
32. Branden C, Tooze J (1999) Introduction to protein structure, 2nd edn. Taylor & Francis Group, London, UK
33. Hamdi M, Ferreira A, Sharma G, Mavroidis C (2008) Microelectronics J 30:190
34. Sivaramakrishnan S, Spink BJ, Sim AYL, Doniach S, Spudich JA (2008) Proc Natl Acad Sci U S A 105:13356–13361
35. Ackbarow T, Sen D, Thaulow C, Buehler MJ (2009) PLoS One 4:e6015-1–e6015-13
36. Ramachandran GN, Sasisekharan V (1968) Adv Protein Chem 23:283–438
37. Ho BK, Thomas A, Brasseur R (2003) Protein Sci 12:2508–2522
38. Zimm B, Bragg J (1959) J Chem Phys 31:526–535
39. Lifson S, Roig A (1961) J Chem Phys 34:1963–1974
40. Enomoto S, Krimm S (1962) Biophys J 2:317–326
41. Cantor C, Schimmel P (1980) Biophysical chemistry. Part III: the behavior of biological macromolecules. WH Freeman, CA
42. Rohs R, Etchebest C, Lavery R (1999) Biophys J 76:2760–2768

43. Hénin J, Schulten K, Chipot C (2006) J Phys Chem B 110:16718–16723
44. Masugata K, Ikai A, Okazaki S (2002) Appl Surf Sci 188:372
45. Navizet I, Cailliez F, Lavery R (2004) Biophys J 87:1426–1435
46. Becker N, Oroudjev E, Mutz S, Cleveland JP, Hansma PK, Hayashi CY, Makarov DE, Hansma HG (2003) Nat Mater 2:278–283
47. Oroudjev E, Soares J, Arcdiacono S, Thompson JB, Fossey SA, Hansma HG (2002) Proc Natl Acad Sci U S A 99(Suppl 2):6460–6465
48. Herrmann H, Bär H, Kreplak L, Strelkov SV, Aebi U (2007) Nat Rev Mol Cell Biol 8:562–573
49. Liu J, Deng Y, Zheng Q, Cheng CS, Kallenbach NR, Lu M (2006) Biochemistry 45:15224–15231
50. Crick FHC (1952) Nature 170:882–883
51. Magin TM, Vijayaraj P, Leube RE (2007) Exp Cell Res 313:2021–2032
52. Pekny M, Lane EB (2007) Exp Cell Res 313:2244–2254
53. Norlén L, Masich S, Goldie KN, Hoenger A (2007) Exp Cell Res 313:2217–2227
54. Kreplak L, Richter K, Aebi U, Herrmann H (2008) Meth Cell Biol 88:273–297
55. Fudge DS, Levy N, Chiu S, Gosline JM (2005) J Exp Biol 208:4613–4625
56. Schopferer M, Bär H, Hochstein B, Sharma S, Mücke N, Herrmann H, Willenbacher N (2009) J Mol Biol 388:133–143
57. Kellermayer MSZ, Karsai A, Kengyel A, Nagy A, Bianco P, Huber T, Kulcsár A, Niedetzky C, Proksch R, Grama L (2006) Biophys J 91:2665–2677
58. Steinbacher S, Seckler R, Miller S, Steipe B, Huber R, Reinemer P (1994) Science 265:383–386
59. Kanamaru S, Leiman PG, Kostyuchenko VA, Chipman PR, Mesyanzhinov VV, Arisaka F, Rossmann MG (2002) Nature 415:553–557
60. Goto Y, Yagi H, Yamaguchi K, Chatani E, Ban T (2008) Curr Pharm Des 14:3205–3218
61. Tombolato L, Novitskaya EE, Chen PY, Sheppard FA, McKittrick J (2009) Acta Biomaterialia 6:319–330
62. Knowles TP, Fitzpatrick AW, Meehan S, Mott HR, Vendruscolo M, Dobson CM, Welland ME (2007) Science 318:1900–1903
63. Gunari N, Walker GC (2008) Langmuir 24:5197–5201
64. Nakajima K, Nishi T (2006) Chem Rec 6:249–258
65. Oesterhelt F, Rief M, Gaub H (1999) New J Phys 1:6.1–6.11
66. Afrin R, Alam MT, Ikai A (2005) Protein Sci 14:1447–1457
67. Marszalek PE, Li H, Oberhauser AF, Fernandez JM (2002) Proc Natl Acad Sci USA 99:4278–4283
68. Rivetti C, Walker C, Bustamante C (1998) J Mol Biol 280:41–59
69. Scholtz JM, Baldwin RL (1992) Annu Rev Biophys Biomol Struct 21:95–118
70. Chakrabartty A, Kortemme T, Baldwin RL (1994) Protein Sci 3:843–852
71. Chakrabartty A, Baldwin RL (1995) Adv Protein Chem 46:141–176
72. Scholtz JM, Qian H, York EJ, Stewart JM, Baldwin RL (1991) Biopolymers 31:1463–1470
73. Lantz MA, Jarvis SP, Tokumoto H, Martynski T, Kusumi T, Nakamura C, Miyake J (1999) Chem Phys Lett 315:61–68
74. Kageshima M, Lantz AM, Jarvis SP, Tokumoto H, Takeda H, Ptak A, Nakamura C, Miyake J (2001) Chem Phys Lett 343:77–82
75. Dixon D, Dobbs K, Valentini JL (1994) J Phys Chem 98:13435–13439
76. Baldwin RL (2003) J Biol Chem 278:17581–17588
77. Idiris A, Alam MT, Ikai A (2000) Protein Eng 13:763–770
78. Doty P, Wada A, Yang JC, Blout ER (1957) J Polymer Sci 23:851
79. Rief M, Pascual J, Saraste M, Gaub HE (1999) J Mol Biol 286:553–561
80. Li L, Wetzel S, Plückthun A, Fernandez JM (2006) Biophys J 90:L30–L32
81. Law R, Carl P, Harper S, Dalhaimer P, Speicher DW, Discher DE (2003) Biophys J 84:533–544
82. Root DD, Yadavalli VK, Forbes JG, Wang K (2006) Biophys J 90:2852–2866
83. Oesterhelt F, Oesterhelt D, Pfeiffer M, Engel A, Gaub HE, Müller DJ (2000) Science 288:143–146

84. Müller DJ, Kessler M, Oesterhelt F, Möller C, Oesterhelt D, Gaub H (2002) Biophys J 83:3578–3588
85. Afrin R, Takahashi I, Shiga K, Ikai A (2009) Biophys J 96:1105–1114
86. Chakrabarti B, Levine AJ (2005) Phys Rev E 71:031905-1–031905-15
87. Chakrabarti B, Levine AJ (2006) Phys Rev E 74:031903-1–031903-11
88. Morozov AV, Kortemme T, Tsemekhman K, Baker D (2004) Proc Natl Acad Sci USA 101:6946–6951
89. Arora N, Jayaram B (1997) J Comput Chem 18:1245–1252
90. Tamashiro MN, Pincus P (2001) Phys Rev E 63:021909-1–021909-8
91. Saito N, Go M (1968) J Phys Soc Jpn 24:376
92. Doi M (1974) Polymer J 6:222–229
93. Courty S, Gornall JL, Terentjev EM (2005) Proc Natl Acad Sci U S A 102:13457–13460
94. Courty S, Gornall JL, Terentjev EM (2006) Biophys J 90:1019–1027
95. Tsuji Y, Yasunaga T, Sano T, Ushio H (1976) J Am Chem Soc 98:813–818
96. Bösterling B, Engel J (1979) Biophys Chem 9:201–209
97. Williams S, Causgrove TP, Gilmanshin R, Fang KS, Callender RH, Woodruff WH, Dyer RB (1996) Biochemistry 35:691–697
98. Thompson PA, Eaton WA, Hofrichter J (1997) Biochemistry 36:9200–9210
99. Clarke DT, Doig AJ, Stapley BJ, Jones GR (1999) Proc Natl Acad Sci U S A 96:7232–7237
100. Huang CY, Klemke JW, Getahun Z, DeGrado WF, Gai F (2001) J Am Chem Soc 123:9235–9238
101. Tegenfeldt JO, Prinz C, Cao H, Chou S, Reisner WW, Riehn R, Wang YM, Cox EC, Sturm JC, Silberzan P, Austin RH (2004) Proc Natl Acad Sci U S A 101:10979–10983
102. Reccius CH, Mannion JT, Cross JD, Craighead HG (2005) Phys Rev Lett 95:268101–268101
103. De Gennes PG (1979) Scaling concepts in polymer physics, 1st edn. Cornell University Press, Ithaca, NY
104. Doi M, Edwards S (1986) The theory of polymer dynamics, 1st edn. Oxford University Press, Oxford, UK
105. Childs PRN (2004) Mechanical design, 2nd edn. Elsevier Butterworth-Heinemann, Amsterdam, The Netherlands, pp 239–240
106. Timoshenko S, Goodier J (1970) Theory of elasticity. London, England, pp 429–432
107. Jalalahmadi J, Naghdabadi V (2007) J Phys Conf Ser 61:497–502
108. Suezaki Y, Go N (1975) Int J Pept Protein Res 7:333–334
109. Thacher T, Rabitz H (1988) Biophys J 54:695–704
110. Yamazaki T, Furuya H, Watanabe T, Miyachi S, Nishiuchi Y, Nishio H, Abe A (2005) Biopolymers 80:225–232
111. Abe A, Furuya H, Okamoto S (1997) Biopolymers 43:405–412
112. Sivaramakrishnan S, Spink BJ, Sim AY, Doniach S, Spudich JA (2008) Proc Natl Acad Sci (USA) 105:13356–13361
113. Spink BJ, Sivaramakrishnan S, Lipfert J, Doniach S, Spudich JA (2008) Nat Struct Mol Biol. 15:591–597

Editor: A. Abe

Adv Polym Sci (2010) 232: 97–207
DOI:10.1007/12_2010_50

Published online: 21 April 2010

Bioactive Polymer/Hydroxyapatite (Nano)composites for Bone Tissue Regeneration

Kinga Pielichowska and Stanislaw Blazewicz

Abstract Bioactive polymer/hydroxyapatite (nano)composites are currently being intensively investigated as materials for promotion of bone tissue regeneration and reconstruction. The advantages of polymeric biomaterials, compared to metallic or ceramic materials, are the ease of manufacturing components having various and complex shapes, reasonable cost, and their ability to possess a wide range of physical and mechanical properties. Additionally, hydroxyapatite (HAp) is one of the most attractive materials for bone implants because of its composition and biological similarity to natural tissues. It can be obtained in a nanostructured form, which facilitates its fine dispersion in the polymer matrix as well as producing advantageous interactions with bioactive polymer and tissue. This paper reviews recent advances in polymer/(nano)HAp composites and nanocomposites for bone tissue regeneration, with particular emphasis on the material characteristics. Specific topics associated with polymer/HAp composition, molecular orientation and morphology, surface modifications, the interactions between the components, and their biological behaviours are described. Finally, the challenges facing this emerging field of research are outlined.

Keywords Biomedical polymers · Hydroxyapatite · Macromolecular nanomaterials · Tissue regeneration

Contents

K. Pielichowska (✉) and S. Blazewicz
AGH University of Science and Technology, Faculty of Materials Science and Ceramics, Department of Biomaterials, Al. Mickiewicza 30, 30-059 Kraków, Poland
e-mail: kingapie@agh.edu.pl; blazew@agh.edu.pl

Abbreviations

ACP	Amorphous calcium phosphate
AFM	Atomic force microscopy
Alg	Alginate
ATRP	Atom transfer radical polymerization
BBC	Bioactive bone cement
BC	Bacterial cellulose
calHAp	Calcined HAp
cdHAp	Calcium-deficient hydroxyapatite, $Ca_{10-x}(HPO_4)_x(PO_4)_{6-x}(OH)_{2-x}$
CF	Carbon fibre
cHAp	Carbonated hydroxyapatite
Col	Collagen
CS	Chitosan
DMA	Dynamic mechanical analysis

DSC	Differential scanning calorimetry
EA	Ethylene-acrylic acid copolymer
ECM	Extracellular matrix
fsHAp	Flame-spheroidized hydroxyapatite
FTIR	Fourier-transform infrared spectroscopy
g-HAp	Grafted HAp
GBR	Guided bone regeneration
Gel	Gelatin
GF/PL	Gas foaming and particulate leaching
HAp	Hydroxyapatite, $Ca_{10}(PO_4)_6(OH)_2$
HDPE	High-density polyethylene
HMWPE	High molecular weight polyethylene
HV	Hydroxyvalerate
HYA	Hyaluronic acid
IPN	Interpenetrated polymer network
MTDSC	Modulated temperature DSC
nanoHAp	Nanosized hydroxyapatite
nsHAp	Non-sintered HAp
PA	Polyamide
PAN	Polyacrylonitrile
PAsp	Poly(aspartic acid)
PC	Polycarbonate
PCL	Poly(ε-caprolactone)
PCLVPA	Poly(ε-caprolactone-*co*-vinyl phosphonic acid) copolymer
PCM	Phase change material
PDLLA	Poly(D,L-lactide)
PE	Polyethylene
PEEK	Poly(ether ether ketone)
PET	Poly(ethylene terephtalate)
PHB	Polyhydroxybutyrate
PHBHV	Poly(hydroxybutyrate-*co*-hydroxyvalerate) copolymer
PHEMA	Poly(hydroxyethyl methacrylate)
PHV	Polyhydroxyvalerate
PLGA	Poly(D,L-lactic-*co*-glycolic acid) copolymer
PLLA	Poly(L-lactide)
PMMA	Poly(methyl methacrylate)
POM	Polyoxymethylene, polyacetal resin
PP	Polypropylene
PPF	Polypropylene fumarate
PSU	Polysulphone
PSZ	Partially stabilized zirconia
PTFE	Poly(tetrafluoroethylene)
PU	Polyurethane
PVC	Poly(vinyl chloride)
SAXS	Small angle X-ray scattering

SBF	Simulated body fluid
SCA	Starch/cellulose acetate
SC/PL	Solvent casting and particulate leaching
SEM	Scanning electron microscopy
semi-IPN	Semi-interpenetrating polymer network
SFF	Solid freeform fabrication
sHAp	Sintered HAp
SLS	Selective laser sintering
SPPSU	Sulphonated polysulphone
stHAp	Stoichiometric HAp
STHAp	Surface-treated HAp
TCP	Tricalcium phosphate, $Ca_3(PO_4)_2$
TEM	Transmission electron microscopy
TG	Thermogravimetry
TTCP	Tetracalcium phosphate, $Ca_4P_2O_9$
ucalHAp	Uncalcined HAp
UHMWPE	Ultrahigh molecular weight polyethylene
WAXD	Wide-angle X-ray scattering
XPS	X-ray photoelectron spectroscopy
XRD	X-ray diffraction

1 Introduction

Bone is one of the most commonly replaced tissues of the human body. The human body can efficiently repair fractures in bone, and small defects are often repaired by the growth of natural bone that is even stronger than the original bone. Large defects generally cause problems because the body is incapable of repairing them. Traditional methods for filling these defects have used acellular cadaver bone or autologous bone, but both have serious limitations as well as a long history [1, 2]. Currently, both biological and synthetic grafts have been used for bone repair. Bone grafting is widely applied in reconstructive bone surgery, and particularly in large and critical-size bone defect arising from trauma, tumours, infections or congenital deficiencies. The procedure with the use of natural bone grafts is traumatic and associated with donor site morbidity in the case of autogenous grafts. In the case of allo- or xenogenic transplants, there is also risk of disease transmission and immune rejection.

The worldwide number of orthopaedic surgerical operations performed is growing steadily due, in part, to the combination of longer average human life span and more active lifestyles. In the USA alone, 152,000 total hip replacements and 299,000 knee replacement operations were undertaken and 59,000 revisions of hip and knee replacements were carried out in 2000, so that over $5 billion is currently being spent annually on orthopaedic related conditions. As these procedures become more popular, because many patients want to continue to lead an active life, the number of such procedures is expected to increase [3].

Tissue engineering introduces a new therapeutic philosophy associated with replacement, restoration, maintenance and enhancement of tissue and organ functions. Although this research area is relatively new, it has already made significant practical progess. At present, it seems that tissue engineering is one of the most influential domains within new strategies for the treatment of diseased tissues and organs. Hence, bone tissue engineering is a new area of research with clinical applications in orthopaedic defects, bone tumours, pseudoarthrosis treatment, stabilization of spinal segments, as well as in maxillofacial, craniofacial, and neck and head surgery. During recent decades, different biomaterials of biological or synthetic origin have been designed with the intention of providing extracellular matrix (ECM) scaffolds for new bone formation [4]. Such scaffolds are three-dimensional (3D) constructions that are implanted into the human body, leading to host integration and acceptance.

Different materials have been investigated for applications in bone tissue engineering – metals, ceramics and polymers. As with all materials implanted into the body, the polymers for bone regeneration must be biocompatible. In addition, the materials should be either mouldable, shapeable, or polymerizable in situ to ensure good integration in the defective area [1]. Tissue engineering scaffolds should facilitate the colonization of cells and possess properties and characteristics that enhance cell attachment, proliferation, migration and expression of native phenotypes. Scaffold characteristics and properties such as porosity, the surface-area-to-volume ratio, pore size, pore interconnectivity, structural strength, shape (or overall geometry) and biocompatibility are crucial factors in the design and fabrication of polymeric materials for bone tissue engineering [5, 6].

In recent years, the search for innovative bone substitutes for developing this new therapeutic concept has concentrated on non-metallic composite materials, particularly on polymer/ceramic composites and nanocomposites. Such (nano)composites consisting of a polymer matrix and bioactive micro/nanofillers provide specific biomaterials for internal bone implants with biological and mechanical properties tailored for a given medical use. These materials link the advantages of polymers (structural stability, strength, biocompatibility, desired shape) with properties of ceramics that resemble those of bone structure.

The bioactivity of these composites is rendered by the composite's bioactive component which, when implanted into the human body, interacts with the surrounding bone and, in some cases, even with the soft tissue. This occurs through a time-dependent modification of the surface, triggered by its implantation within the living bone. An ion-exchange reaction between the bioactive implant and the surrounding body fluids results in the formation of a biologically active carbonated hydroxyapatite (cHAp) layer on the implant that is chemically and crystallographically equivalent to the mineral phase of bone. Prime examples of bioactive materials are natural or synthetic hydroxyapatite (HAp, $Ca_{10}(PO_4)_6(OH)_2$), tricalcium phosphate, glass-ceramic A-W, and Bioglass [7]. The polymer matrix provides the ductility and other associated properties that are required for hard tissue replacement materials [8].

One of the most important groups of polymer/ceramic composites are polymer/HAp materials. HAp is a well-known bioceramics that is chemically similar to human inorganic hard tissue [9]. The ECM of all bone tissue is essentially built by mineralized collagen (Col) fibrils, consisting of a Col matrix reinforced with 50–60 wt% of apatite crystals. HAp is the closest synthetic equivalent to human bone mineral and it exhibits excellent biocompatibility, bioactivity, bone-bonding properties and osteoinduction, as shown by in vivo studies [10]. Due to the chemical similarity, HAp facilitates the formation of new bone without resorption and interacts with the living system. If HAp is in a nanostructured form, the interaction with the lowest hierarchical levels of a bone's structure becomes even more intensive because of the material's high surface area and enhanced bioactivity. To date, a number of polymer/(nano)HAp composites have been engineered to mimic important aspects of the structure and properties of human bone tissue.

In this review, recent advances in polymer/(nano)HAp composites and nanocomposites for bone tissue regeneration are presented, including specific subjects associated with polymer/HAp composition, molecular orientation and morphology, surface modification, and interactions between components and the biological environment.

2 Bone Structure

Bone is a vascularized, dense, supporting skeletal tissue consisting of cells and mineralized ECM [2]. The bone matrix consists of polymeric core (collagen type I, ColI) and calcium phosphate in the form of HAp. A cortical bone layer (compact bone) forms the outer region of long bones, while trabecular bone (cancellous bone) fills the interior. The major component of compact bone is an osteon – it creates cylindrical conduits known as Haversian canals, which provide access for the circulatory and nervous system [1].

Bone is deposited by bone-forming cells (osteoblasts) and by osteocytes, some of which are ciliated. Osteoblasts cease dividing when they transform into osteocytes. Bone is moulded, remoulded and/or removed by mono- or multinucleated osteclasts (and sometimes by osteocytes). The first bone matrix deposited is unmineralized and known as osteoid. Subsequently, the osteoid is impregnated with HAp to form bone – the mineralized tissue [2].

From the biomaterials engineering viewpoint, human bone tissue is a composite comprized of a Col matrix reinforced with 40–50 vol% apatite crystals. The apatite crystals are plate-like and elongated with a *c*-axis-preferred orientation in the direction of principal stress, such as the longitudinal anatomical axis of long bones. It follows that bone tissue exhibits anisotropic mechanical properties. The elastic moduli of human cortical bone in the longitudinal and transverse directions are typically reported as being in the range of 16–23 and 6–13 GPa, respectively, and depend on the person's age [11, 12].

The mechanical properties of bone are related to its complex hierarchical structure. Several levels of structural organization, from macro- to subnanostructure, can be identified: (a) the macrostructure (cancellous and cortical bone); (b) the microstructure (from 10 to 500 mm: Haversian systems, osteons, single trabeculae); (c) the submicrostructure (1–10 mm: lamellae); (d) the nanostructure (from a few hundred nanometers to 1 mm: fibrillar Col and embedded mineral); and (e) the subnanostructure (below a few hundred nanometers: the molecular structure of constituent elements, such as mineral, collagen, and non-collagenous organic proteins). The hierarchical structural organization of bone is shown in Fig. 1 [12].

This structure has an irregular, yet optimized, arrangement and orientation of the components, accounting for the heterogeneous and anisotropic properties of the bone material [12].

Currently, most attention is being paid to the subnanostructure of bone tissue, which is believed to govern bone's unique biological and mechanical properties (Fig. 2).

The most prominent structures of bone tissue at this scale are the Col fibres, which are surrounded and infiltrated by the mineral. Mineral crystals, Col fibres and non-collagenous organic proteins are the three main materials of bone subnanostructures.

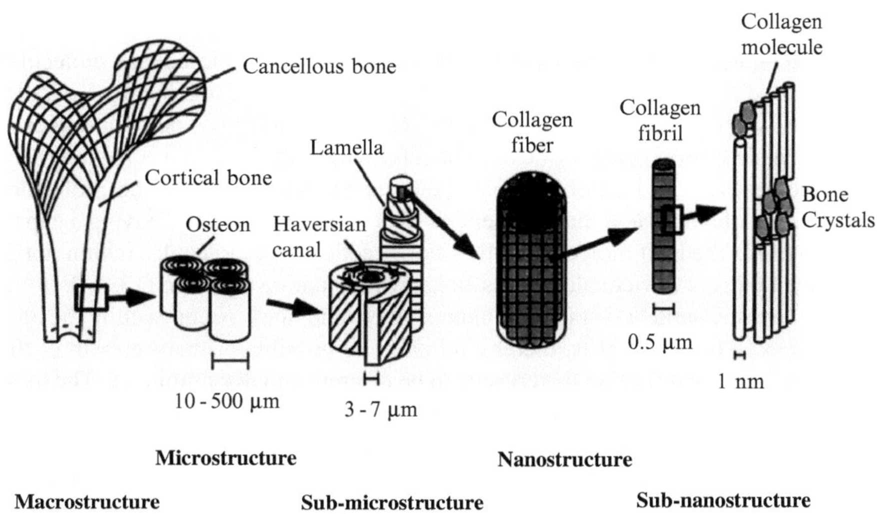

Fig. 1 Hierarchical structural organization of bone: (*left to right*) cortical and cancellous bone; osteons with Haversian systems; lamellae; collagen fibre assemblies of collagen fibrils; bone mineral crystals, collagen molecules and non-collagenous proteins. Reprinted from [12] with permission from Elsevier

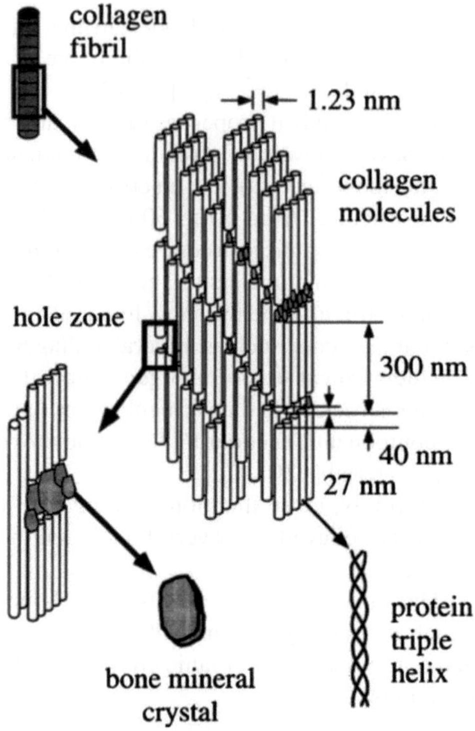

Fig. 2 Assembly of collagen fibrils and fibres and bone mineral crystals. The well-known 67-nm periodic pattern results from the resence of adjacent hole (40 nm) and overlap (27 nm) regions of the assembled molecules. Reprinted from [12] with permission from Elsevier

ColI molecules are triple helices with a length of about 300 nm. After excretion, the globular ends are cleaved off by enzymes and the triple-helical molecules undergo a self-assembly process leading to a staggered arrangement of parallel molecules, with a periodicity *D*, of 67 nm. Gap regions appear as a consequence of this staggered arrangement of Col molecules within fibrils because the length of the molecules is not an integer multiple of the staggered period *D*. Hence, molecules have a length of a little less than 5*D* periods (5×67 nm $= 335$ nm), leaving a gap of about 35 nm to the next molecule in the axial direction. Col molecules within fibrils are joined by just a few covalent cross-links, which mature with age [13, 14].

The mature apatite crystals are plate-shaped and they occur within the discrete spaces in the Col fibrils, thereby limiting the possible primary growth of the mineral crystals, and forcing the crystals to be discrete and discontinuous. The mineral crystals grow with a specific crystalline orientation – the *c*-axes of the crystals are parallel to the long axes of the Col fibrils and the average size of plates are 50×25 nm and 2–3 nm thick. The bone apatite has limited amounts of impurities such as HPO_4, Na, Mg, K, citrate, carbonate and others, whose positions and configurations are not completely known [12].

The high toughness of bone results from the ability of its micro- and nanostructure to absorb the deformation energy without crack nucleation and propagation [15, 16]. Both mineral nanoparticles and the mineralized fibrils initially deform, but to different degrees [17].

Nanomechanical heterogeneity is expected to influence the elasticity, damage, fracture and remodelling of bone. Complementary nanomechanical investigations of the local stiffness and hardness of the osteon reveal a modification of mechanical properties at the lamellar level [18]. Specifically, the compressive modulus of the sublamellae within a single lamella, as measured by nanoindentation, varies from about 17 to 23 GPa, with thin layers of lower stiffness alternating with wider layers of higher stiffness [13].

The spatial distribution of nanomechanical properties of bone is quantified by the length of individual Col fibrils. The elaborated patterns of stiffness, ranging from ~2 to 30 GPa, which do not correlate directly with the topographical features, are attributed to underlying local structural and compositional variations. A new energy-dissipation mechanism arising from nanomechanical heterogeneity, which offers a mean of enhancing ductility, damage evolution and toughening, has been proposed and is supported by computational simulations that incorporate the nanoscale experimental results. These simulations predicted that non-uniform inelastic deformation over larger areas and increased energy dissipation arising from nanoscale heterogeneity lead to markedly different biomechanical properties compared to a uniform material [19].

The knowledge about bone tissue structure and morphology described in detail in [2, 13, 20–25] proves to be very complex because bone is arranged in several hierarchical structures. Rational design of artificial implant materials should take into account the tissue's characteristics – the more similarities possessed by the fabricated implant to the bone tissue (biomimetism), the greater the chance of acceptance of the alien system by the human organism.

3 Hydroxyapatite

HAp is one of the most important bioceramics for medical and dental applications because it is the main inorganic constituent of human hard tissue like bone and teeth. The chemical structure of HAp is presented in Fig. 3.

HAp, together with other materials of the calcium phosphate family, such as β-tricalcium phosphate (β-TCP, $Ca_3(PO_4)_2$) and tetracalcium phosphate (TTCP, $Ca_4P_2O_9$), belongs to group of bioactive ceramic materials. Most of the biological apatites are non-stoichiometric, poorly crystalline, and contain several foreign ions (mainly carbonate $CO_3{}^{2-}$) and traces of Na^+, Mg^{2+}, Fe^{2+}, Cl^-, $HPO_4{}^{2-}$ and F^-. Among them, $CO_3{}^{2-}$ ions play a vital role in bone metabolism because they comprise about 3–8 wt% of the calcified tissue which may vary depending on the bone age [26–28]. Essentially, bone mineral is a calcium-deficient HAp (cdHAp) with a Ca:P ratio of about 1.5; it is chemically and compositionally similar to β-TCP, while being structurally similar to stechiometric HAp (Ca:P = 1.67). Apatites with Ca:P ratios ranging from 1.67 to 1.33 originate from the loss of Ca^{2+} ions from the unit cell (hence the name calcium-deficient hydroxyapatites) and have the formula $Ca_{10-x}(HPO_4)_x(PO_4)_{6-x}(OH)_{2-x}$ [8, 29].

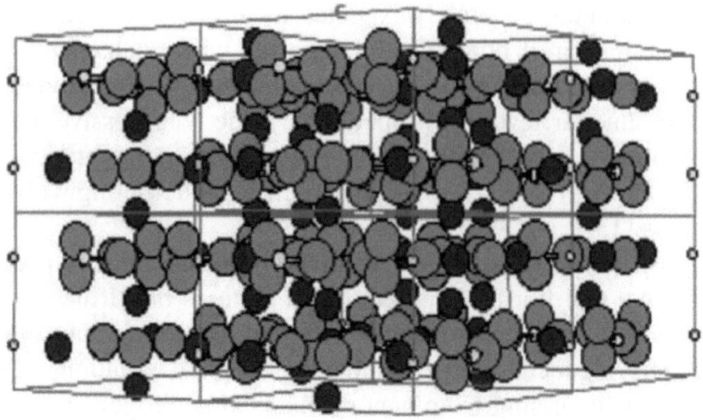

Fig. 3 Atomic structure of HAp; smallest *white* atoms, phosphorus; largest *gray* atoms, oxygen; medium *black* atoms, calcium. Reprinted from [22] with permission from Elsevier

HAp can be obtained from natural sources (animal bones or corals) or synthesized from appropriate substrates, including carbonates.

Various methods have been used to synthesize HAp [26–36]. For instance, Hao et al. presented results of HAp nanocrystal synthesis by the hydrothermal method [30, 31]. In this procedure, $(NH_4)_2HPO_4$ aqueous solution is slowly dropped into a stirred aqueous $Ca(NO_3)_2$ solution at room temperature. The resultant precipitate is put into an autoclave and hydrothermally treated at 140°C for 5 h under a pressure of 0.3 MPa, followed by centrifugal washing with deionized water. The hydrothermal slurry is finally dried to produce HAp as a white powder [31].

In mature bone, HAp nanocrystals are all irregularly shaped thin plates of carbonate apatite with average length and width of 50 × 25 nm and thickness of 2–3 nm. Compared with sintered HAp (sHAp) particles, the hydrothermally formed needle-like nanocrystals possess greater similarity to the morphology and crystal structure of natural apatite [30].

Carbonated hydroxyapatite (cHAp) powders were obtained by Slosarczyk et al. using a wet method [35, 36]. In their syntheses, CaO, $Ca(NO_3)_2{\cdot}4H_2O$ or $Ca(CH_3COO)_2{\cdot}H_2O$ were used as the calcium reagents; H_3PO_4 or $(NH_4)_2HPO_4$ were the sources of phosphorous (the Ca:P molar ratio was 1.67); and NH_4HCO_3 or $NaHCO_3$were applied as reactants introducing CO_3^{2-} groups. Biological apatites present in natural bone, dentin and enamel contain different amounts of carbonate: 7.4, 5.6 and 3.5 wt%, respectively. In synthetic powders prepared by this method some fraction of the PO_4^{3-} and OH^- groups are replaced by CO_3^{2-} groups. Among the variety of HAp-based bioceramics, cHAp seems to be a promising material for bioresorbable bone substitution [35].

HAp single-crystal nanowires were synthesized by Zhang et al. by template technology, and characterized using X-ray diffraction (XRD), transmission electron microscopy (TEM), and X-ray photoelectron spectroscopy (XPS). The results confirmed that the crystalline order of the HAp precursors was retained in the electrodeposited nanowires. The HAp single-crystal nanowires, growing in *c*-axis coorientation along the direction of the template, displayed structural similarity to the HAp found in natural bone [37].

Rapid synthesis of cdHAp with Ca:P ratio 1.5 was performed by Siddharthan and coworkers [29] by a precipitation method using calcium nitrate tetrahydrate/phosphoric acid under microwave irradiation. Microwaves were used to accelerate the formation of cdHAp with a shorter processing time than other available methods. Characterization of the cdHAp confirmed the nanocrystalline nature of the synthesized powder, which consisted of needle-like nanocrystallites, 16–39 nm long and 7–16 nm thick. In another work, Murugan et al. [27] applied microwaves for the synthesis of cHAp. The overall results indicated that cHAp has structural and chemical functionalities quite similar to biological apatite. The in vitro solubility of cHAp under physiological conditions was found to be higher than HAp, which is a good indication of its bioresorbable nature.

Zhou et al. [38] investigated the nanoemulsion technique for synthese of cHAp nanospheres for tissue engineering scaffolds. Nanoemulsions, similar to microemulsions, are a new class of emulsions having very fine and uniform droplet sizes, typically in the range of 20–200 nm [39]. Nanoemulsions offer the possibility of obtaining microemulsion-like dispersions without the need to use high surfactant concentrations, or even without using any surfactant. The nanoemulsion technique has been found to be very promising for the synthesis of nanosized, B-type cHAp particles of spherical shape [38].

Recently, in the work of Landi et al. [40], synthetic HAp doped with Mg was prepared by a wet-chemical synthesis. The best results were obtained for 5.7 mol% Mg-doped HAp – it was biocompatible since it showed no genotoxicity, carcinogenicity or cytotoxicity. Moreover, the authors found that HAp-Mg granulate was comparable to, or even better than, traditional HAp as a bone substitute when tested in vivo; in particular, HAp-Mg showed greater osteoconductivity over time and a higher material resorption than stoichiometric HAp (stHAp).

Generally, HAp possesses excellent biocompatibility and is osteoconductive. It has been used clinically as a bioactive bone substitute in the form of powder, porous matrix or dense body and as a bioactive coating on total hip prostheses [8, 29].

HAp-based implants are made both from natural products, such as animal bones or corals, and from synthetic HAp. HAp materials manufactured from animal bones, e.g. bovine or pig bones, have the advantage since they inherit some properties of the raw material, such as its chemical composition and structure [41–43]. HAp-based implants from synthetic HAp are preferable because of their uniform composition, high biocompatibility, overall safety and the fact that their microstructure is completely controllable [44].

However, HAp is difficult to shape in the complex forms required for bone treatment because of its hardness and brittleness. HAp powders, used for the treatment of bone defects, have the limitation that they easily migrate from the implanted sites. Therefore, novel composites of HAp and organic polymers that can compensate for the poor mechanical properties of HAp have become of great interest [45]. The concept of using bioactive HAp particles/modified polymer composites as implant materials for bone replacement was introduced by Bonfield in the early 1980s [46]. Since then, HAp-modified polymer-based biocomposites have been widely studied for bone tissue replacement. They are described in detail in the forthcoming chapters of this review.

4 Polymers in Bone Tissue Engineering

Metallic materials, such as stainless steel and cobalt–chromium alloys, have been used as bone implants in orthopaedic areas for a long time. However, these materials can experience stress-shielding problems due to their high elastic modulus [47]. If a stiff metal or ceramic implant is placed in bone, the bone will be subjected to a reduced mechanical environment, and consequently it will resorb. In the case of total hip replacement, bone resorption in the proximal femur that leads to aseptic loosening of the prosthesis (which is a very common problem) is believed to be caused by the state of stress and strain in the femoral cortex after the metallic femoral hip replacement is implanted [8]. Other major disadvantages of metallic and ceramic materials for tissue engineering applications are the lack of degradability in a biological environment and their very limited processability [48–51]. Compared to metallic or ceramic materials, the advantages of polymeric biomaterials, which have been introduced into clinical use since the 1960s [52, 53], are the ease of manufacturing of products with various and complex shapes, reasonable cost and their availability with a wide range of physical and mechanical properties. For certain applications, it may also be an advantage that the stiffness of polymeric materials is much closer to the stiffness of bone in contrast to metals or ceramics [54, 55]. Tensile strength and the elastic modulus of polymers make it possible to apply them as bone implants (Fig. 4).

On the basis of long-lasting clinical programmes, the most important synthetic nondegradable polymers in bone tissue engineering are: polyethylene (PE),

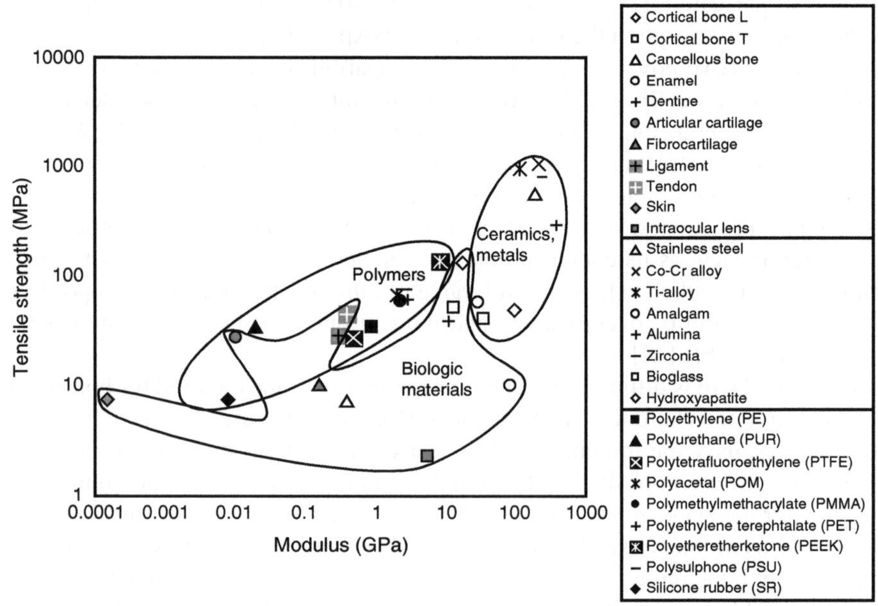

Fig. 4 Tensile strength vs. Young's modulus of materials with relevance for composite design when considering biomedical applications. Reprinted from [56] with permission from Elsevier

polypropylene (PP), selected polyurethanes (PUs), polytetrafluoroethylene (PTFE), poly(vinyl chloride) (PVC), polyamides (PAs), poly(methyl methacrylate) (PMMA), polyoxymethylene (POM, polyacetal resin), polycarbonate (PC), poly(ethylene terephtalate) (PET), poly(ether ether ketone) (PEEK), and polysulphone (PSU). These polymers are also considered to be biostable in the human body and have found wide applications in the medical field, ranging from PTFE vascular grafts to ultrahigh molecular weight polyethylene (UHMWPE) acetabular cups [8, 56, 57].

Another group of polymers that have attracted increasing attention for their use as tissue engineering scaffolds in the last decade are degradable polymers, in particular polyesters (e.g. polylactide, polyglycolide), proteins and polysaccharides.

The meaning and definition of the words biodegradable, bioerodable, bioresorbable and bioabsorbable, which are often used misleadingly in the tissue engineering literature, are of primary importance in discussing the rationale, function and chemical and physical properties of polymer-based scaffolds [58]. In this paper, the biorelated polymer properties are based on the definitions given by Vert et al. [58, 59]:

> Biodegradable are solid polymeric materials and devices which break down due to macromolecular degradation with dispersion in vivo but there is no proof for the elimination from the body (this definition excludes environmental, fungi or bacterial degradation). Biodegradable polymeric systems or devices can be attacked by biological elements so that the integrity of the system and in some cases (but not necessarily) of the macromolecules themselves, is affected and gives fragments or other degradation by-products. Such fragments can move away from their site of action but not necessarily from the body.
>
> Bioresorbable are solid polymeric materials and devices which show bulk degradation and further resorb in vivo; i.e. polymers which are eliminated through natural pathways either because of simple filtration of degradation by-products or after their metabolisation.
>
> Bioresorption is thus a concept which reflects total elimination of the initial foreign material and of bulk degradation by-products (low molecular weight compounds) with no residual side effects. The use of the word bioresorbable assumes that elimination is shown conclusively.
>
> Bioerodable are solid polymeric materials or devices, which show surface degradation and further, resorb in vivo. Bioerosion is thus a concept, too, which reflects total elimination of the initial foreign material and of surface degradation by-products (low molecular weight compounds) with no residual side effects.
>
> Bioabsorbable are solid polymeric materials or devices, which can dissolve in body fluids without any polymer chain cleavage or molecular mass decrease. For example, it is the case of slow dissolution of water-soluble implants in body fluids. A bioabsorbable polymer can be bioresorbable if the dispersed macromolecules are excreted.

Biodegradation of polymeric biomaterials involves cleavage of hydrolytically or enzymatically sensitive bonds in the polymer, leading to polymer erosion. Depending on the mode of degradation, polymeric biomaterials can be further classified into hydrolytically degradable polymers and enzymatically degradable polymers. Most of the naturally occurring polymers undergo enzymatic degradation [60].

Moreover, depending on the mechanism of degradation, polymeric biomaterials can be classified into bioresorbable and bioerodable. Polymer erosion can be of two types, bulk erosion and surface erosion. In ideal bulk erosion (also termed homogeneous erosion), material is lost from the entire polymer volume at the same time due to water penetrating the bulk; bulk-eroding polymers degrade all over their cross-section and have erosion kinetics that are nonlinear and are usually characterized by a discontinuity [61]. In surface erosion (also termed heterogeneous erosion), material is lost from the polymer surface only and the erosion rate will be proportional to the surface area. These are generally hydrophobic polymers or polymers containing a hydrophobic component wherein water cannot penetrate easily into the bulk [62, 63, 67].

Degradable materials have already demonstrated promising results in clinical use, for example as resorbable surgical sutures and meshes or in bone tissue engineering [53, 56, 60, 65, 66]. However, a number of problems have been encountered regarding the use of these polymers in tissue engineering applications due to the release of acidic degradation products, leading to inflammatory responses. Another limitation of biodegradable polymers is that they lack a bioactive function, e.g., for bone tissue applications, they do not allow for bone apposition or bonding on the polymer surface [67].

No single polymer can meet all the requirements for bone tissue engineering scaffolds. Each polymer material has its own characteristic advantages and disadvantages. Composites and nanocomposites can offer a suitable set of properties, which often show an excellent balance between strength and toughness, and usually possess improved characteristics compared to their individual components [68]. Since natural bone is an organic/inorganic hybrid material made of collagen and apatite, composites consisting of a polymer matrix and HAp reinforcement seem to be suitable candidates for bone tissue engineering applications [48].

5 Polymer/HAp (Nano)composites in Bone Tissue Engineering

It is well-known that two crucial factors in producing (nano)composites with bone-like properties are (a) a good interfacial adhesion between organic polymers and inorganic HAp, and (b) the uniform dispersion of HAp at nanolevel in the polymer matrix [69]. Lack of adhesion between HAp particles and polymer matrix usually results in early failure at the interface of both phases. On the other hand, if HAp is not uniformly dispersed in polymer matrix through proper processing methods it might agglomerate and the resulting composite material is characterized by poor mechanical properties. In the last decade, considerable effort was made to produce and characterize polymer/nanoHAp systems that resemble bone structure to the high extent that classical microcomposites do. Two main groups of polymers are used in fabrication of polymer/HAp composites in bone tissue engineering – nondegradable and degradable polymers. Both groups consist of a number of polymers, and are described consecutively in Sects. 6 and 7.

6 Nondegradable Polymer/HAp (Nano)composites

Nondegradable polymers are widely used in bone tissue engineering due to good mechanical properties and chemical stability. The desired advances are primarily related to improving their biocompatibility and performance, both of which are already remarkable in terms of actual clinical applications. Nondegradable/HAp systems are usually applied when tissue cannot be regenerated due to its large losses or for elderly patients with a less effective self-healing ability of the tissue.

Nondegradable polymers that are used as polymer matrices for HAp are presented in Table 1.

6.1 *Polyethylene*

Polyethylene (PE) as a polymer matrix for (nano)composites with HAp is mostly applied in its high-density (HDPE) or UHMWPE form. UHMWPEs have very high wear resistance, chemical resistance and low coefficient of friction, and are self-lubricating because of the linear character of the chains and their high molecular weight of 2–10 million. In powder form, they are processed by sintering, compression moulding or extrusion [70]. However, the limited lifetime of UHMWPE is still one of the major problems in, e.g. total hip replacement. Efforts to solve the wear problem have resulted in numerous studies on the improvement of UHMWPE by ceramic reinforcements, high-temperature recrystallization or crosslinking [55].

PE is a ductile polymer that allows the incorporation of large amounts of bioceramic particles in the system. Even with a high content of bioceramic particles, the polymer can still be melt-processed using current technologies. Particulate HAp-reinforced HDPE composites have been developed since the early 1980s by Bonfield [46] as analogue materials for bone replacement. They were commercialized (HAPEX) by Smith & Nephew Richards in 1995 and used for production of middle ear implants. They were the first bioactive ceramic/polymer composites that were designed to mimic the structure and to the match properties of bone, and have given rise to the research and development of other bioactive composites using the same rationale [8]. The close elastic modulus matching of HDPE/HAp composite to bone shows promise in solving the problem of bone resorption that has been encountered with the use of implants made up of conventional materials, such as metals and ceramics, which possess much higher modulus values than human cortical bone [71].

Various aspects of the mechanical performance of HDPE/HAp composites have been investigated. Wang [8] found that by varying the amount of HAp in the composite, a range of mechanical properties can be obtained. An increase in the volume percentage (vol%) of HDPE in the HAp leads to increases in the Young's modulus, shear modulus, storage modulus (in the dynamic mechanical analysis), microhardness, and tensile strength of HDPE/HAp, with corresponding decreases in the strain to fracture and energy for impact fracture. The particle morphology and average

Table 1 Characteristics of nondegradable polymer/HAp (nano)composites

Polymer	Polymer/HAp (nano)composite	Average size of HAp particles	HAp content	Young's modulus (GPa)	Tensile strength (MPa)	Comments	Ref.
Polyethylenes	HDPE/HAp	4.14 μm 7.32 μm	10–45 vol%	0.65–5.54 0.72–5.39	17–21 17–21	HAPEX™; blended, compounded, centrifugal milled and compression moulded	[72]
	HDPE/HAp	4.53 μm 7.43 μm	10–45 vol%	1.97–6.43	n/a	Compression moulded	[73]
	HMWPE/HAp	4.5 μm	10–50 wt%	3–7.5	50–90	Bi-axial rotating drum processing and twin screw extrusion	[78]
	HDPE/HAp	n/a	7–40 vol%	0.88–4.29	17–23	Silanation of HAp and acrylic acid grafting of PE	[71]
	HDPE/HAp	n/a	5–40 vol%	n/a	n/a	–	[74]

(continued)

Table 1 (continued)

Polymer	Polymer/HAp (nano)composite	Average size of HAp particles	HAp content	Young's modulus (GPa)	Tensile strength (MPa)	Comments	Ref.
	HDPE/HAp	4.0 μm 3.6 μm	40 vol%	n/a	n/a	The effect of morphological features of HAp on the rheology and processability	[77]
	HDPE/HAp	18 × 2.3 μm 1.3 μm	10–60 vol%	2.5–10.8 2.2–8.3	16–30 10–26	Compression moulded, whiskers and powder	[79]
	HDPE/HAp	10.1 μm	10 wt%	n/a	26–30	Treatments based on silane chemicals	[76]
	HDPE/HAp HDPE/HAp/PSZ	n/a	26–69 wt%	n/a	n/a	The effect of partially stabilized zirconia (PSZ) on the biological properties	[80]

(continued)

Table 1 (continued)

Polymer	Polymer/HAp (nano)composite	Average size of HAp particles	HAp content	Young's modulus (GPa)	Tensile strength (MPa)	Comments	Ref.
	UHMWPE/HAp	2 μm	14–50 wt%	6.8	27	In paraffin oil using twin-screw extrusion and then compression moulded	[82]
	UHMWPE/HAp	2 μm	50 vol%	4.1	n/a	Swelling/twin-screw extrusion, compression moulded, and then hot drawn	[83]
	HDPE/HAp MDPE/HAp LDPE/HAp	n/a	30%	1.3 1.0 0.6	28 22 9	Compounded in the extruder, composite preforms are compression moulded and irradiated (cross-linked)	[84]
Polyamides	PA8063/HAp	10–30 nm × 50–90 nm	39–65 wt%	n/a	n/a	Co-solution method	[87]
	PA-6,6/HAp	20–30nm × 60–100 nm	50 wt%	n/a	39–42	Co-solution method	[86]

(continued)

Table 1 (continued)

Polymer	Polymer/HAp (nano)composite	Average size of HAp particles	HAp content	Young's modulus (GPa)	Tensile strength (MPa)	Comments	Ref.
	PA-6,6/HAp	10–20 nm × 70–90 nm	39–65 wt%	n/a	65–87	Needle-like nanoHAp, prepared by co-solution, co-precipitation method and water treatment	[88]
	PA-6,6/HAp	n/a	35–55 wt%	n/a	n/a	Co-solution method, thermal and crystalline behaviour	[90]
	PA-6/HAp	10–30 nm × 70–90 nm	65%	n/a	n/a	Co-solution method	[89]
Polyacetal	POM/HAp	99% <100 nm	0.5–5 wt%	0.9–1.1	n/a	Twin-screw extrusion, injection moulded	[100]
Polysulphones	PSU/HAp	7.32 μm	5–10 vol%	n/a	n/a	Blended, compounded, compression/ injection moulded	[102]

(continued)

Table 1 (continued)

Polymer	Polymer/HAp (nano)composite	Average size of HAp particles	HAp content	Young's modulus (GPa)	Tensile strength (MPa)	Comments	Ref.
	SPPSU/HAp	n/a	9 wt%	n/a	n/a	Sulphonated polysulphone (SPPSU) spontaneous precipitation of HAp in aqueous suspensions of the polymer particles	[103]
	PSU/HAp	n/a	n/a	2.8–3.9	n/a	Pre-preg method	[104]
	PSU/HAp	n/a	15 wt%	n/a	n/a	Injection moulded	[105]
Poly(ether ether ketone)	PEEK/HAp	n/a	10–90 wt%	n/a	n/a	Thermal analysis	[110]
	PEEK/HAp	90 wt% below 60 μm	10–40 wt%	n/a	n/a	Selective laser sintering	[5]
	PEEK/HAp	25.8 μm	0–40 vol%	2.8–16.0	45–69	Flame-spheroidized HAp, melt compounded, granulated and injection moulded	[109]

(continued)

Table 1 (continued)

Polymer	Polymer/HAp (nano)composite	Average size of HAp particles	HAp content	Young's modulus (GPa)	Tensile strength (MPa)	Comments	Ref.
	PEEK/HAp	19.94 μm	5–40 vol%	2.5–13	50–100	Melt compounded, granulated and injection-moulded	[108]
	PEEK/HAp	n/a	10–40 vol%	n/a	n/a		[107]
	PEEK/HAp	21.6 μm × 2.8 μm	0–50 vol%	7.3–23.2	42–90	Whisker reinforcements, compression moulded	[11]
Poly(methyl methacrylate)s	PMMA/HAp						
	PBMA/HAp	n/a	n/a	n/a	n/a	Surface grafted, bone cement	[122]
	PHEMA/HAp						
	PMMA/HAp	n/a	20 wt%	n/a	n/a	Bone cement	[125]
	PMMA/HAp	n/a	4.6–8.8 vol%	n/a	n/a	Bone cement	[113]
	PMMA/ACP	n/a	40 wt%	n/a	n/a	Internal conversion of ACP into HAp	[126]
	PMMA/CS/HAp	n/a	40–60 wt%	n/a	n/a	Bone cement, with chitosan	[118]

particle size of HAp were found to affect the mechanical properties of HDPE/HAp composites. In subsequent work by this research team, HDPE composites with synthetic HAp particles of two median sizes and different morphologies (HAPEX) were studied. An increase in HAp volume fraction leads to an increase in both strength and modulus of composites, with a simultaneous reduction in strain to failure. The particle size and morphology of HAp have significant effects on the mechanical properties of HAPEX. HAp particles of large median sizes reduce the strength and modulus, but increase the ductility of composites [72].

The effects of the HAp volume fraction, temperature and HAp particle size on the storage modulus (E') and damping (tan δ) in the HDPE composite were investigated in the subsequent work of Wang's group [73]. They reported that an increase in HAp volume fraction in the composites increased the storage modulus at all temperatures; an increase in temperature reduced the storage modulus values for all volume fractions; and the HAp with smaller particle size reinforced PE more efficiently. Moreover, an increase in relative modulus with increasing temperature was observed for composites of higher HAp volume fraction. This implies the existence of a thermally stressed region in the filled PE matrix near the HAp particles. These thermally induced stresses reduced the modulus and restricted the damping behaviour of the matrix near the HAp particles, hence reducing the damping at lower temperatures. Storage moduli obtained by dynamic mechanical analysis (DMA) were consistently higher than, but linearly related to, Young's moduli obtained by quasi-static tensile tests. The authors proposed four damping regions in investigated composites (shown schematically in Fig. 5): (1) the filler (HAp), which does not contribute to damping; (2) the PE/HAp interface; (3) the interphase, i.e., the matrix region near the interface, which is under thermally induced stresses; and (4) the bulk PE matrix away from the filler.

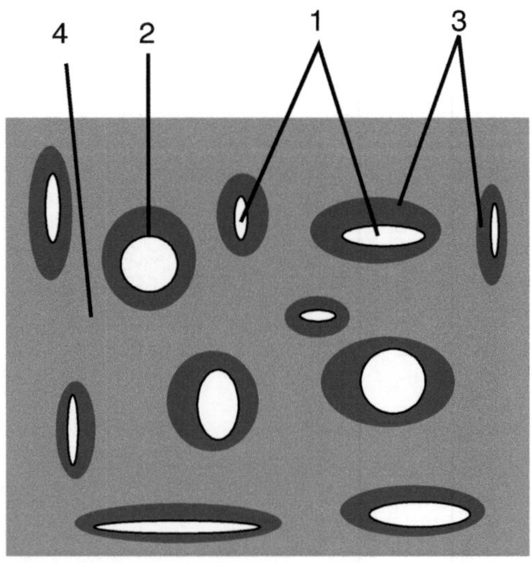

Fig. 5 Thermally stressed PE around HAp particles. *1* HAp filler, *2* PE–HAp interphase, *3* interface under thermally induced stresses, *4* bulk PE matrix. Reprinted from [73] with permission from Springer

The regional variations in the matrix suggest that the damping mechanisms for the filled and unfilled PEs are dissimilar at low temperatures, but similar at higher temperatures. The reduction in the thermal stresses in the PE around the HAp particles allows the damping behaviour of the filled polymers to be closer to that of the unfilled polymer [73].

Tribological properties of unfilled HDPE and HDPE/HAp composites against the duplex stainless steel in dry and lubricated conditions (distilled water or aqueous solutions of proteins) were also investigated [74]. Generally, HDPE/HAp composites had lower coefficients of friction than unfilled HDPE under certain conditions. HDPE underwent severe deformation and damage at high sliding speeds and hence was not suitable for implants, which may be subjected to extreme conditions. For HDPE/HAp composites, the lower testing speed induced severe degradative reactions as compared to the higher testing speed in the presence of aqueous proteins. The incorporation of HAp at up to 10 vol% improves both the friction and wear behaviour of HDPE, but the addition of HAp above 10 vol% deteriorates the tribological properties of the composites due to cavitational effect of lubricants and corrosive reactions of proteins.

Because the mechanical properties of HDPE/HAp composites manufactured via an established conventional route approach the lower limit for human cortical bone [75] and the load-bearing application of this composite material is limited, silanation of HAp and acrylic acid grafting of PE to improve bonding between HAp and PE were applied [71]. As a result, a silicon-containing interphase between the HAp particle and the PE matrix was formed. Use of the silane coupling agent also facilitated the penetration of polymer into cavities in individual ceramic particles, which resulted in enhanced mechanical interlocking at the matrix-reinforcement interface. Due to a strong bond (both mechanical and chemical) being formed between HAp and PE, chemically coupled HDPE/HAp composites possess improved mechanical properties and fracture behaviour [71].

In the parallel study, Sousa and coworkers [76] investigated several coupling treatments based on silanes for the development of HDPE/HAp composites with enhanced interphase properties. Two HAp powders, sintered (sHAp) and non-sintered HAp (nsHAp) were used in combination with silanes, namely methacryloxypropyl trimethoxy silane (MEMO), 3-(2-aminoethyl)aminopropyl trimethoxy silane (DAMO), vinyl trimethoxy silane (VTMO), 3-aminopropyl triethoxy silane (AMEO) and trimethoxypropyl silane (PTMO). For the sHAp composites, the most effective coupling treatments concerning stiffness were those based on MEMO and AMEO. For the nsHAp composites, it was possible to improve significantly both the stiffness and the strength. Amino silanes demonstrated to be highly efficient concerning strength enhancement. The higher effectiveness of the coupling treatments for nsHAp filled composites was attributed by the authors to higher particle surface area, smaller particle size distribution and expected higher chemical reactivity. For both cases, the improvement in mechanical performance after the coupling treatment was consistent with the enhancement in interfacial adhesion observed by scanning electron microscopy (SEM) in the form of characteristic morphology changes [76].

Elsewhere, Joseph and coworkers [77] studied the effect of surface area and morphology of HAp on the rheology and processability of injection-moulding grade HDPE/HAp composites. The results demonstrated that the surface morphology of HAp has a considerable influence on controlling the rheology and processability of the HDPE/HAp materials. They explained this phenomenon as a result of the larger surface area of HAp – it needs more matrix to wet the surface, thus making less matrix available for shear flow. Lowering the specific surface area of HAp in composites reduces the shear viscosity and energy for processing. The authors suggested that by using low surface area HAp that can be produced by sintering, production of polymer/HAp composites can be further optimized, depending on the application [77].

The macromolecular-orientation approach for the production of HMWPE and HAp ductile composites was applied by Reis et al. [78]. The composites were produced with different amounts (10–50% by weight) of the reinforcement by two procedures: bi-axial rotating drum and twin screw extrusion. The final processing stage was conventional injection moulding or injection moulding with shear-controlled orientation under a wide range of processing conditions. The obtained PE/HAp composites were ductile, even for 50 wt% of HAp. The use of injection moulding with shear-controlled orientation was a unique way of inducing anisotropy to thick sections and to produce very stiff composites that could be used in biomedical applications with high mechanical loads. This fact, combined with the bioactive behaviour of the HAp phase, made the composite produced in the course of the study usable for orthopaedic load-bearing implants [78].

In another approach, synthetic HAp whiskers were utilized as reinforcement for orthopaedic biomaterials [79]. High-density polyethylene (HDPE) was reinforced with either the synthesized HAp whiskers or a commercially available spherical HAp powder using a novel powder processing technique that facilitated uniform dispersion of the additive in the matrix prior to compression moulding. An increase of the volume fraction of either reinforcement type over 0–50 vol% resulted in increased elastic modulus, a maximum in ultimate tensile stress and decreased work to failure. Due to alignment of the whiskers in the matrix during processing, composites containing HAp whiskers were anisotropic and had higher elastic modulus, ultimate tensile strength and work to failure relative to composites reinforced with spherical HAp (Fig. 6).

In another work, Yari Sadi and coworkers [80] studied the effect of partially stabilized zirconia (PSZ) on the biological properties of the HDPE/HAp composites by investigating the simultaneous effect of HAp and PSZ volume fractions on the in vitro response of human osteoblast cells. It was found that the volume fraction of HAp has a significant effect on the bioactivity of the composites. The composites provided a favourable site for cell attachment, with cells frequently observed anchoring to the HAp particles. Interestingly, the results show that the addition of PSZ into the HDPE/HAp composites does not adversely affect the biological properties of these composites and, in some cases, composites with PSZ showed better biological results than HDPE/HAp systems [80].

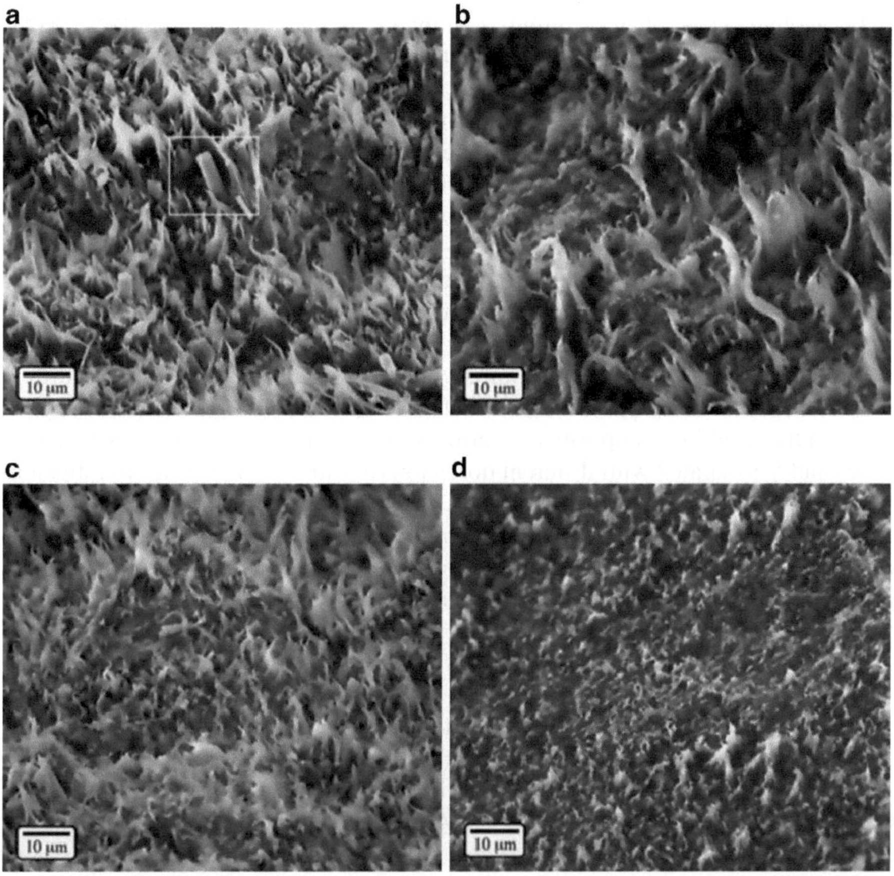

Fig. 6 SEM micrographs showing the failure surface of HDPE composites reinforced with (**a**) 20 vol% HAp whiskers, (**b**) 20 vol% spherical HAp, (**c**) 50 vol% HAp whiskers, and (**d**) 50 vol% spherical HAp. Reprinted from [79] with permission from Wiley InterScience

Recently, UHMWPE has attracted considerable attention as a matrix for polymer/HAp composites. Fang et al. [81, 82] obtained UHMWPE/HAp nanocomposites by compounding UHMWPE and HAp mixtures in paraffin oil using twin-screw extrusion and then compression moulding. SEM images revealed that HAp aggregates were broken down to nanosized particles and homogeneously dispersed in UHMWPE by the combined processes of UHMWPE swelling treatment and twin-screw extrusion. Swelling of the UHMWPE increased chain mobility, thus allowing HAp particles to penetrate the fibril interspaces during twin-screw extrusion. The composite with the HAp volume fraction of 23% exhibited a Young's modulus nine times bigger than that of UHMWPE, while the composite maintained the excellent toughness feature of UHMWPE. The fracture strain reached over 300%, significantly higher than for other types of biocomposites [82].

In another work, Fang's group continued preparation of UHMWPE/HAp nanocomposites with a HAp volume fraction of 50% by combined swelling/twin-screw extrusion, compression moulding, and then hot drawing [83]. The authors reported that HAp nanoparticles were homogeneously dispersed in the UHMWPE matrix and formed an interpenetrated network (IPN); moreover, the UHMWPE fibrils in the composites were highly oriented along the hot drawing direction. The hot drawn composite with draw ratio of 15 exhibited a yield strength of 100 ± 22 MPa, which was comparable to that of cortical bone. The biocomposites demonstrated an excellent ability to induce calcium phosphate formation in simulated body fluid (SBF), which was promising for further applications as load-bearing bone substitutes.

Smolko and Romero [84] examined composites prepared using PE of different densities and HAp powder. Raw materials were compounded in an extruder and then the resulting composite pre-forms were compression moulded into desired plates and γ-irradiated with different doses to cross-link the material. Investigators observed that with HAp incorporated in the PE matrix, an increase in the values of Young's modulus (stiffness) occurred, while elongation at break decreased with the amount of filler, showing the increase of brittleness. Tensile stresses at yield and at break decreased with the filler content for low-density PE (LDPE) and middle-density PE (MDPE) and stayed constant for HDPE. Moreover, the increase in radiation dose improved strength values and substantially reduced the percentage elongation.

Albano and coworkers [85] investigated the improvement of the composite's properties by means of the addition of surface-treated and untreated HAp (STHAp and HAp, respectively). An ethylene-acrylic acid (EA) copolymer was melt-blended with HDPE and HAp (HDPE/HAp/EA). A surface treatment was performed using either an EA copolymer or acrylic acid. Incorporation of EA copolymer as well as STHAp had a remarkable effect on the mechanical properties of composites. The crystallinity degree as well as the crystallization and melting temperatures did not vary. Regarding in vitro evaluation, composites with HAp and EA copolymer had a better cell-adhesion response than those without copolymer. The results of the STHAp composites could be attributed to the electrostatic interactions taking place between the acrylate groups, in EA copolymers and acrylic acid, and the polar groups of the HAp [85].

6.2 Polyamides

Polyamides (PAs) are an important group of semicrystalline engineering polymers that offer a wide spectrum of properties and processing characteristics. Among different PAs, polyamide 6,6 (PA-6,6) has good mechanical properties, compatibility with the human body and is considered to be a useful polymer for biomedical applications. PA-6,6 displays some structural similarity to collagen, and the depolymerization products (hexamethylenediamine and hexanedioic acid) act in vivo as antibacterial agents [86]. An important drawback is associated with the fact that PA-6,6 generally does not show specific interactions with bone tissue, so modification with ceramic additives is a suitable method for preparation of PA-6,6-based biomaterials. This approach was applied by Huang and coworkers, who obtained PA-6,6/HAp nanocomposites [86] by the solvent casting method. The authors used nanosized HAp (nanoHAp) in the form of needle-like crystals using a hydrothermal method. The crystals had a narrow particle size distribution and crystal structure similar to bone apatite. NanoHAp dispersed uniformly in PA-6,6 matrix causes a reinforcement effect and can prevent micro-crackle spreading into cleft as well as fracture during the deformation process. Mechanical testing showed that the PA-6,6/nanoHAp composites have good mechanical properties and may be regarded as promising bone replacement materials.

In a parallel study by Wei et al., polyamide (PA 8063) and nanoHAp were used to prepare bone-like composites [87]. The authors suggested that the direct use of nanoHAp slurry and PA solution to prepare the nanocomposites by the co-solution method is a very useful way to make bioactive composites with high HAp content, good homogeneity and chemical interface bonding. When compared to natural bone, the synthetic nanocomposites can be said to be a bone-like biomaterial. The nanoHAp provides the bioactivity for the composite through interactions with natural bone, and PA is responsible for mechanical strength and toughness. Therefore, PA/HAp nanocomposites may be one of the better groups of bioactive materials for load-bearing bone repair or substitution [87].

Wei et al. [88] found that tissue engineering scaffold material of PA-6,6/nanoHAp composite prepared by the co-solution/co-precipitation method had good homogeneity and high HAp content. Interface chemical bondings are formed between the nanoHAp and PA-6,6 in the composite, which has properties close to that of natural bone.

In subsequent work by this group, PA-6/nanoapatite (nanoA) biocomposites, obtained by a co-solution method, were described [89]. The nanoA of 10–30 nm diameter and 70–90 nm length was uniformly distributed in the PA-6 matrix to form the nanocomposite. The authors reported that molecular interactions and chemical bonding existed between nanoA and PA-6, which greatly improved the mechanical properties and integrity of the composite. They also found that the composite with a high nanoA content (around 65%) had mechanical strength similar to natural bone. In another experiment, an interconnected porous PA-6/nanoHAp material with porosity of 80% and mean pore size of approximately 300 μm was prepared by an injecting foam method. When implanted into cortical bone, the composite combined directly with the natural bone without fibrous capsule tissue between implant and host bone, thus proving its biocompatibility.

Zhang and coworkers studied the thermal stability and crystallization behaviour of PA-6,6 and PA-6,6/nano-HAp biocomposites with reference to nanoHAp content using thermogravimetry (TG) and differential scanning calorimetry (DSC) methods [90]. Investigators found that the thermal stability of the PA-6,6/nano-HAp composites was higher than that of PA-6,6, which was due to the hydrogen bonds between the amide group of PA-6,6 and hydroxyl group of nanoHAp. DSC investigations showed that the incorporation of nanoHAp into PA-6,6 caused an apparent increase in the crystallization temperature and a decrease in the degree of crystallinity. The latter effect can be explained by the role of nanoadditive during the polymer's crystallization process – at higher concentrations it might hinder formation of regular arrangements.

6.3 Polyacetals

The term "polyacetal" is commonly used in the area of biomaterials and it means "polyoxymethylene homopolymer" (hPOM) or copolymer of trioxane with dioxolane or ethylene oxide (also denoted as cPOM). In this paper, the names "polyacetal" and "POM" will be used alternatively. Polyoxymethylene has the general formula $(–OCH_2–)_n$ and its linear backbone structure is similar to that of PE. Properties that arise from this structure are: high-impact strength, stiffness, resiliency, toughness and high yield stress; low friction coefficient; low gas and vapour permeability; exceptional dimensional stability and dielectric properties; high fatigue strength; and good retention of properties at elevated temperature [91]. POM shows an ultrahigh Young's modulus, which is caused by the presence of extended-chain crystals, whereby a part of the polymer could still appear as chain-folded lamellae. Extended chains can be obtained in topotactic polymerization or during processing, when the melting region is approached and a cooperative process throughout crystals becomes dominating [92]. The POM copolymer usually contains cyclic ether segments (e.g. ethylene oxide), which provide increased resistance to unzipping degradation of the polymer chain. The copolymer form of polyacetal has slightly superior resistance to unzipping to monomer (formaldehyde), but its mechanical properties are generally inferior to those of the homopolymer [93].

Polyacetal has been used in different joints, but also as heart valve replacements. Although it is a well-known engineering polymer and has been applied previously in orthopaedic implants (e.g., in hip and knee prostheses), there is little data about the effect of long-term exposure to the physiological environment on its mechanical properties [94, 95]. Moore et al. [96], suggested the possibility of using polyacetal instead of metal for the femoral component of knee prosthesis. One post-mortem specimen, retrieved 9 years after surgery, showed no measurable polyacetal wear and negligible HDPE wear. The histology of tissue in contact with polyacetal was indistinguishable from that adjacent to polymethyl methacrylate and UHMWPE in the same knee. Polyacetal homopolymer has also been used for nearly 30 years as the occluder disc of the Björk–Shiley Delrin (BSD) heart valve [97]. Fourier transform infrared spectroscopy (FTIR), viscometry and chromatography of controls and implants retrieved after up to 20 years all indicated that no systematic degradation of polyacetal took place under exposure to body fluids. Moreover, POM has been used for more than two decades in dentistry as a substitute for acrylic resins and metals in many prosthetic applications. Cytotoxicity studies and intracutaneous reactivity studies of pedodontic POM crowns were performed at NAMSA (Northwood, OH), and the results showed, under the conditions of these studies, no evidence of cytotoxicity or allergenic reactions [98].

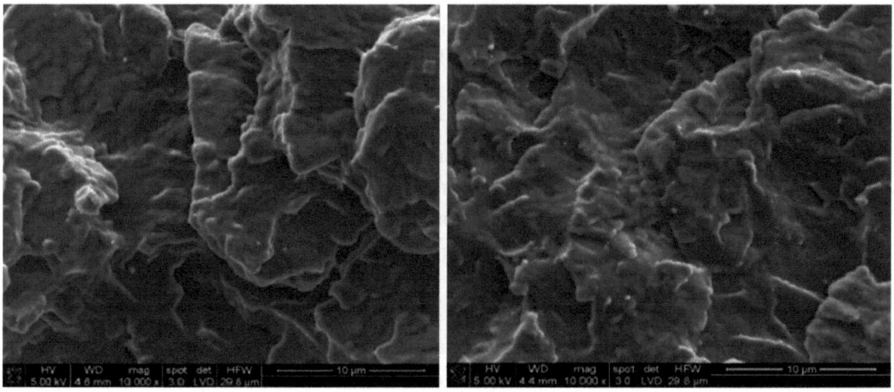

Fig. 7 SEM micrographs of POM (*left*) and POM/5% HAp (*right*)

New bioactive POM/HAp nanocomposites for bone long-term implants have been obtained by Pielichowska using the double-screw extruder. The method ensures homogeneous distribution of the filler within the POM matrix in core–shell architectures (Fig. 7).

The influence of processing conditions on the mechanical and thermal properties of the nanocomposites obtained was investigated. It has been found, based on the DSC results for the first heating cycle, that for pure POM after extrusion the change of the heat of melting is larger than for virgin POM or for POM after injection moulding. This effect can be associated with prevailing directional alignment of the macromolecules during the extrusion process. For POM/HAp nanocomposites the influence of the extrusion's conditions diminishes: HAp nanoparticles act as effective nucleating sites and the rate of crystallization is higher at the initial stages of the process, leaving less chance for crystallization during flow-oriented arrangement of macromolecules [99, 100]. Mechanical properties of the POM/HAp nanocomposites are better than those of POM because HAp causes an increase in the degree of crystallinity of the polymer matrix, and the nanocomposites are bioactive in SBF [101].

6.4 Polysulphone

Polysulphone (PSU) is an amorphous polymer characterized by the presence of sulphone groups in its repeating units. Most commercially important PSUs are aromatic and may contain ether linkages to enhance oxidation resistance. As a class of polymers, PSUs show excellent thermal and mechanical stability, low creep, good dielectric properties, transparency, self-extinguishing properties and resistance to chemicals. The most important and popular PSUs are polyarylsulfone and polyethersulfone [70]. To develop bioactive composites for load-bearing prostheses, PSU may be a better choice for the matrix of a composite than HDPE as its strength and modulus are significantly higher, which can provide a higher level of mechanical properties for composites. Furthermore, PSU has high resistance to β- and γ-radiation and can be steam-sterilized.

Wang and coworkers studied PSU/HAp composites as new hard tissue replacement materials [102]. Composites were manufactured via a standardized procedure, which included drying, blending, compounding and injection/compression moulding. The authors found that HAp particles were well dispersed in the PSU matrix and that different PSU/HAp compositions could be processed. The glass transition temperature (T_g) of the polymer was not considerably affected by the incorporation of HAp, and water absorption of PSU in PSU/HAp composites reaches an equilibrium after 7 days of immersion in distilled water – water uptake was up to 0.4 wt%. Both PSU and PSU/HAp composites exhibited pseudoplastic flow behaviour: viscosity and shear stress decreased with an increase in temperature, and increasing the HAp content in the composite led to increases in viscosity and shear stress. Moreover, the storage modulus of the composite increased with an increase in HAp content, while tan δ was maintained at nearly the same level below T_g of PSU. After 7 days of immersion in distilled water, the storage modulus of PSU/HAp composite was decreased to less than that of unfilled PSU. With an increase in HAp content up to 40 vol%, the stiffness of PSU/HAp composites also increased and was within the lower limit for bone. Just as for HDPE/HAp composites in biaxial fatigue testing, the torsional stress significantly reduced the fatigue life of PSU composites with HAp [8].

In a parallel study, a biocompatible composite made of sulphonated PSU and HAp was prepared and characterized. The incorporation of HAp was achieved by spontaneous precipitation of HAp at controlled conditions of pH and temperature. The time of the precipitation process was used to regulate the content of inorganic component in the composite. The composite material prepared in this experiment contained 9 wt% of HAp and 91 wt% of polymer. Analysis of the kinetic results of the crystallization process showed that the growth of HAp on the film-shaped composite material followed a polynuclear (nuclei above nuclei) crystal growth mechanism [103].

Elsewhere, composites and nanocomposites of PSU with HAp, both natural (nanoparticles) and synthetic (microparticles), were investigated [104]. Results of investigations showed that addition of HAp to PSU causes an increase in Young's modulus and limitation of creep, but a decrease in tensile strength. The best mechanical parameters were observed for the nanocomposite obtained by using the pre-preg method [104]. The authors suggested that HAp build-up in the artificial biological environment is more intense for composites with nanoparticles of natural HAp, because nanoparticles constitute active centres for apatite secretion.

In efforts to understand the long-term behaviour of PSU/HAp composites under mechanical stress, Chlopek and coworkers prepared and characterized a series of composites containing 15 wt% of natural origin (beef bone) HAp. The investigations were carried out on paddle-shaped samples received by injection moulding. Results obtained during in vitro creep tests enabled estimation of the long-term behaviour of the examined samples under mechanical stress, as well as the effects of biological environment on this behaviour. Biostable PSU/HAp composites can be used as long-term implants at stress levels between 10 and 20% of initial strength [105].

6.5 Poly(ether ether ketone)

Special attention of researchers dealing with biomaterials has been focused on poly(ether ether ketone) (PEEK). PEEK is a polyaromatic, semicrystalline, rigid, thermoplastic polymer with excellent mechanical properties and high resistance to various chemicals and radiation. Moreover, PEEK is bioinert and possesses additional benefits, such as good chemical and fatigue resistance, high temperature durability, high wear properties, high ease of processability with the capacity to be shaped readily by machining and heat contouring to fit the shape of bone. Besides, PEEK is non-cytotoxic and can be repeatedly sterilized using conventional steam, γ-irradiation and ethylene oxide treatment without deterioration of its mechanical properties. All these benefits have rendered particulate reinforced PEEK attractive for biomedical applications, including use as material for orthopaedic implants [47, 106].

The Young's modulus and the tensile strength of injection-moulded PEEK/HAp composites were reported to be in the range of 3–10 GPa and 40–90 MPa, respectively [106]. These results indicate that the mechanical properties of PEEK/HAp composites are in the region of cortical bone, making them a potential candidate for use in load-bearing applications [107].

Hence, Yu et al. [107] investigated the kinetics of growth of apatite layers on PEEK/HAp composites with different HAp volume fractions during immersing in SBF. The examined composites were prepared by homogeneous mixing of PEEK and HAp powders, compaction and sintering. The bioactivity of PEEK/HAp composites with 10, 20, 30 and 40 vol% of HAp was evaluated by immersing the composite disks in SBF for up to 4 weeks. The surface of composite with 40 vol% of HAp was covered by bone-like apatite layer after just 3 days of immersion, while composite with 10 vol% of HAp was coated after 28 days. The growth rate constant for apatite crystals increased with HAp volume fraction of the composite, suggesting that the bioactivity of the PEEK/HAp composites increases with increasing HAp volume fraction. Pure PEEK itself is an inert material and shows no significant change on its surface after immersing for 28 days in SBF; however, after incorporating the bioactive HAp into PEEK a gradual apatite formation was observed.

PEEK was reinforced (up to 50 vol%) with HAp whiskers using a novel powder processing and compression moulding technique [11]. An increase in HAp reinforcement amount resulted in increased elastic modulus, but also in decreased ultimate tensile strength, strain-to-failure and work-to-failure. Composites reinforced with 40–50 vol% of HAp whiskers exhibited elastic moduli similar to that of human cortical bone in the longitudinal direction. Composites with 10 and 20 vol% of HAp whiskers displayed tensile strengths similar to that of human cortical bone (longitudinal direction). The application of compression moulding technique resulted in a preferred orientation of HAp whiskers in the direction of viscous flow. Additionally, a degree of elastic anisotropy similar to that of human cortical bone was observed in PEEK with 40 vol% of HAp load. The authors concluded that PEEK reinforced with HAp whiskers is well-suited for orthopaedic implants, in which tailored, bone-mimetic mechanical properties and bioactivity are desired.

Abu Bakar et al. [108] incorporated HAp particulates into PEEK matrix through a series of processing stages involving melt compounding, granulating and injection moulding. They demonstrated the feasibility of fabricating PEEK composites containing high amounts of HAp using injection moulding. This was achieved through appropriate selection of the processing parameters that match the rheological behaviour of the processed system.

In another study, flames-pheroidized HAp (fsHAp) was incorporated into semicrystalline PEEK by melt compounding, granulating and injection moulding to obtain biocomposites with high HAp loading of up to 40 vol% [109]. Investigators revealed that the tensile modulus and microhardness increased, while strength and strain to fracture decreased correspondingly, with progressive addition of HAp particles. The composite system under investigation also exhibited mechanical properties matching those of human bone With HAp loading beyond 30 vol%, the moduli were within the range of the human cortical bone.

Another approach can be found in the work of Fan and coworkers [47]. To design PEEK/HAp composites with desired mechanical properties, they applied a 3D finite-element unit cell model. The model consists of an elastic-brittle HAp spherical particle, an elasto-plastic matrix and an interphase layer between the particle and the matrix. The authors compared the predicted results with the experimental data obtained from existing literature. The predicted results of the matrix degradation and the effects of the damage and failure on the interphase layer are in good agreement with experimental data. The authors claimed that the finite element model, with appropriate material behaviours of the interphase layer and with consideration of the matrix degradation, is successful in predicting not only the trend of the changes in the mechanical properties of the PEEK/HAp composites in relation to the HAp content, but also in foreseeing the nonlinear constitutive relations.

Tan et al. [5] obtained 3D scaffolds using selective laser sintering (SLS) of PEEK/HAp biocomposites. Blends of PEEK/HAp were produced by mixing pure PEEK and HAp powders in different weight percentages using a roller-mixer; the prepared powder blends were then processed using an SLS system (Fig. 8).

The SLS technique was highly advantageous as it provides good user control over the microstructures of created scaffolds by adjusting the SLS process parameters. Besides, the ability to incorporate different amounts of HAp into the polymer matrix reiterated its viability for use in tissue engineering scaffolds, especially bone scaffolds [5].

Reports on the detailed thermal behaviour of PEEK/HAp composites [as well as other polymer/HAp (nano)composites] are scarce in the literature. Advanced thermal analysis methods, e.g., modulated temperature differential scanning calorimetry (MTDSC) or hyphenated thermoanalytical methods such as thermogravimetry coupled with Fourier transform infrared spectroscopy (TG-FTIR) or mass spectrometry

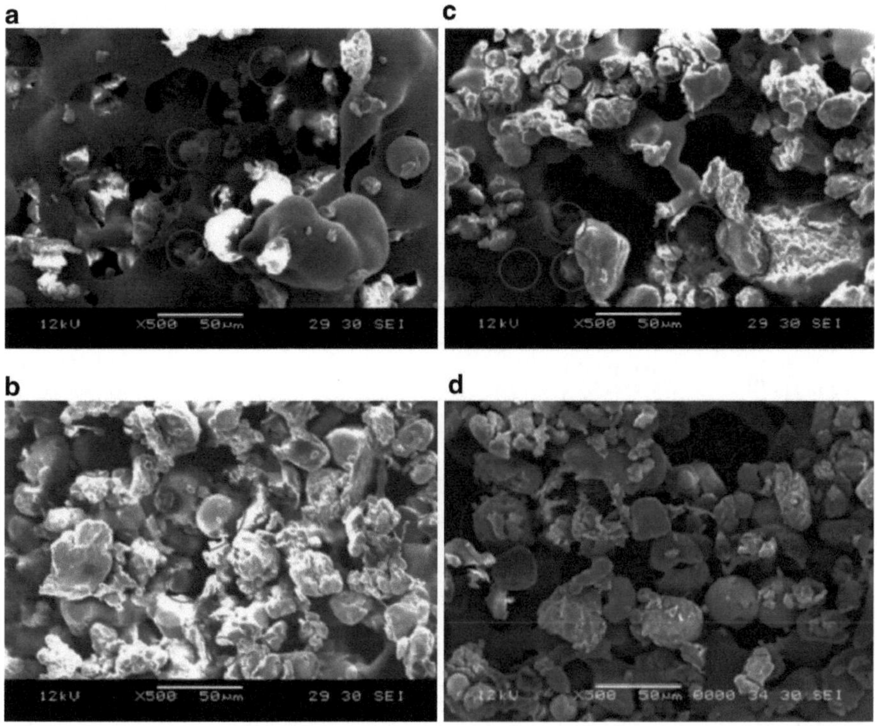

Fig. 8 SEM micrographs of sintered PEEK blend with (**a**) 10 wt% HAp, (**b**) 20 wt% HAp, (**c**) 30 wt%, and (**d**) 40 wt% HAp (HAp particles are *circled*). Reprinted from [5] with permission from Elsevier

(TG-MS), are applied quite rarely, although they may yield useful information for both fabrication (by thermal processing methods) and life-time predictions of polymeric biocomposites.

In the work of Meenan and coworkers [110], the thermal characterization of a series of PEEK/HAp composites was done by TG, DSC and MTDSC. The latter technique offers extended temperature profile capabilities by, e.g., sinusoidal wave superimposed to the normal linear temperature ramp. The capabilities of the MTDSC method in comparison with conventional DSC include separation of reversing (thermodynamic) and nonreversing (kinetic) components of the heat flow, improved resolution of closely occurring and overlapping transitions, and increased sensitivity of heat capacity measurements [111].

The authors found that an increase of HAp content in mixtures with PEEK causes a decrease of the onset temperature of PEEK degradation by up to 13°C. Water loss in the temperature range of 80–120°C and a slow dehydroxylation of the HAp, as the temperature increases over the range 200–750°C, were observed. Using MTDSC, the glass temperature (Tg) as well as the crystallization and melting temperatures (T_c and T_m, respectively) of PEEK/HAp composites were accurately determined from the reversing component of the heat flow (Fig. 9) [110].

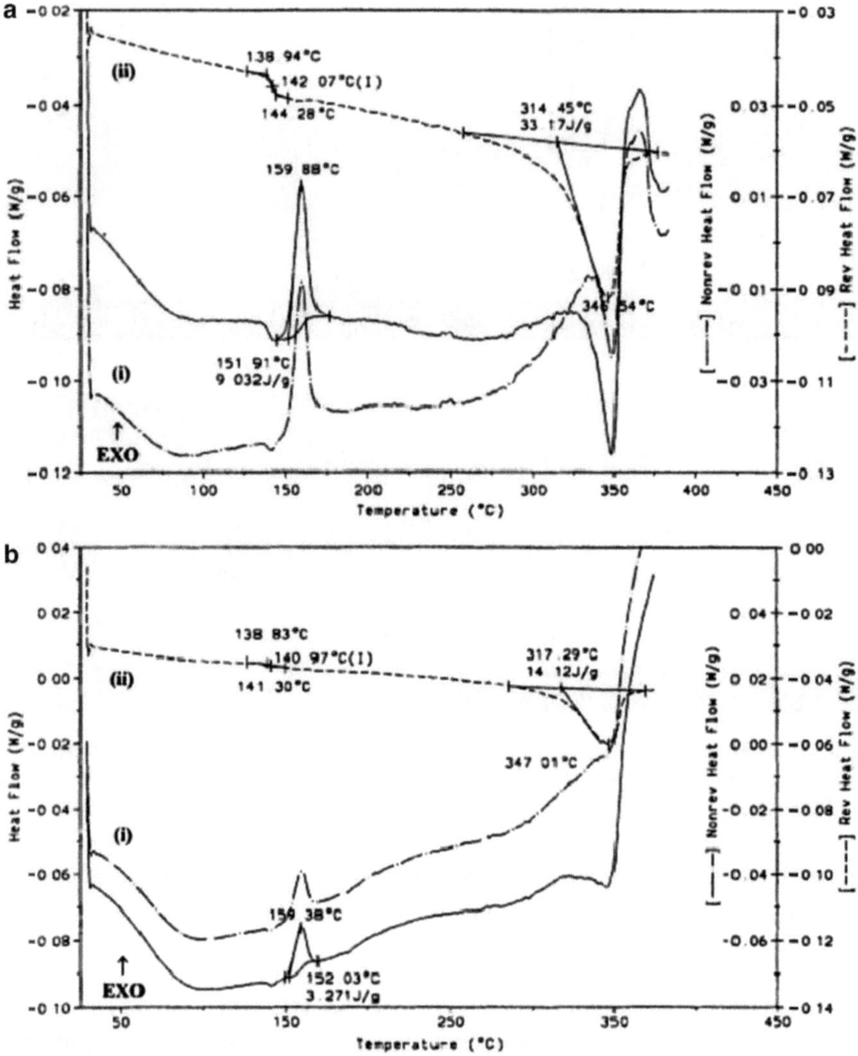

Fig. 9 Modulated temperature DSC scans for (**a**) 74/26% PEEK/HAp with $T_g \approx 142°C$, $T_c \approx 160°C$ and $T_m \approx 347°C$ and (**b**) 37/63% PEEK/HAp with $T_g \approx 141°C$, $T_c \approx 160°C$, $T_m \approx 347°C$. In both plots the total heat flow signal (*solid line*) has been split into (*i*) the non-reversing (kinetic) and (*ii*) reversing (heat capacity) components. Reprinted from [110] with permission from Springer

6.6 Poly(methyl methacrylate)

Poly(methyl methacrylate) (PMMA) and its derivatives are the most commonly used polymers in bone cements for the fixation in orthopaedic surgeries. PMMA is an amorphous thermoplastic polymer that offers excellent optical clarity, resistance to weathering, surface hardness, good chemical resistance, rigidity, dimensional stability, low mould shrinkage, good impact strength, nontoxicity and tastelessness. PMMA powders can be injected and compression moulded, extruded, etc. Liquids can be cast into sheets, lenses and rods [112, 113]. PMMA conforms to the shape of its surroundings, allows even distribution of implant loads, and forms a strong mechanical bond with implants, although mechanical characteristics are rather poor. Polymers produced by mixing of the cement phases are brittle and have a short fatigue lifetime. Widespread use of PMMA is limited by other properties; for instance, PMMA adheres insufficiently to bone surfaces (no bioactivity) [114, 115] and undergoes an exothermic polymerization reaction, releasing heat that has been shown to damage bone tissue [116–118].

To minimize this problem, De Santis and coworkers [119] prepared a blend of PMMA and paraffin-based phase change material (PCM) that is able to store thermal energy [120] produced during polymerization of methyl methacrylate.

Other acryl polymers used for orthopaedic applications include poly(*n*-butyl methacrylate) (PBMA), characterized by lower exothermic effect, higher fracture toughness and superior fatigue life as well as lower toxicity to soft tissue and dental pulp [121], and poly(hydroxyethyl methacrylate) (PHEMA), which shows enhanced biocompatibility [122].

Although PMMA is still the current standard for cementheld prostheses, it is an inert material for fibroblastic cells observed at the bone–cement interface. It forms a strong bond with the implant, but the bond between the cement and the bone is considered to be weak, with fibroblastic cells observed at the implant site. Incorporation of HAp increases the biological response to the cement from tissue around the implant site, thus giving increased bone apposition. Research revealed that the addition of up to 40 wt% of HAp to PMMA cement has been shown to increase the fracture toughness, and that the addition of up to 15 wt% of HAp led to an increase in flexural modulus, while the tensile and compressive strengths remained constant [113].

Dalby et al. [112] used an in vitro tissue culture model to evaluate the biological response of conventional PMMA and PMMA/HAp composites. An increase in biological activity of examined bone cements in response to increasing HAp content was observed and the authors suggested that loading of HAp into cements could be the way forward in producing materials with the flow properties that surgeons require, and the biological properties that benefit the patient.

In another work, Kwon and coworkers conducted studies that showed that a PMMA/HAp composite (30 wt% of HAp) increased the interfacial shear strength at the bone–implant interface 6 weeks after implantation in rabbits [123]. Several other studies also found that the addition of HAp can enhance the mechanical properties of bone cements, although the extent of improvement varied, depending on the type of bone cement [124].

Moursi et al. [125] studied the response of osteoblasts to PMMA/HAp materials. They found that osteoblasts attached equally well to PMMA/HAp and to traditional materials (titanium and PMMA). This attachment was followed by modest early proliferation as compared to control, but similar to that of the other materials.

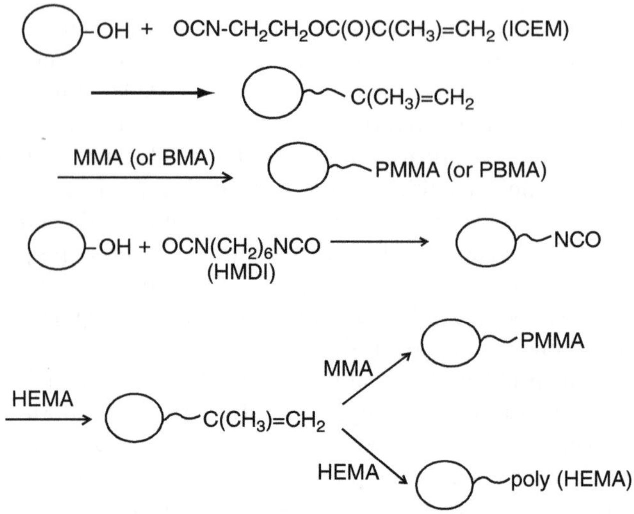

Scheme 1 Two schemes for grafting PMMA, PBMA or PHEMA to the surface of HAp by using either ICEM or HMDI-HEMA. Reprinted from [122] with permission from Wiley InterScience

In contrast, subsequent proliferation of osteoblasts on PMMA/HAp was significantly enhanced (compared to PMMA) after 8 days in culture.

In order to improve adhesion between polymer matrix and HAp, PMMA, PBMA and PHEMA were grafted onto the surface of HAp by using isocyanatoethyl methacrylate (ICEM), or hexamethylene diisocyanate (HMDI)/hydroxyethyl methacrylate (HEMA) This process was investigated by Liu et al. [122] (Scheme 1).

The authors established that all three methacrylate polymers were successfully coupled to HAp particles via covalent bonding of isocyanate groups. The possibility of realizing a chemical bonding between polymer matrix and HAp provides a wide range of possibilities for chemical integration of HAp in polymer composites.

More recently, Skrtic et al. [126] studied methacrylate conversion and volumetric contraction in photopolymerized composites with amorphous calcium phosphate (ACP), which was used because of its relatively high solubility in aqueous media and its rapid transformation to HAp. Ideally, on polymerization, the resin systems used in composites should achieve the contradictory goals of both high vinyl conversion and minimal volumetric contraction. A high degree of vinyl conversion, with homogeneous network formation, is desirable because it maximises the glass transition temperature of the matrix; thereby rendering it less susceptible to the softening effects of the aqueous oral environment. Researchers investigated the degree of vinyl conversion and volumetric contraction of several types of resin composites formulated with three types of ACP. They observed that the conversion of methacrylate functional groups in the resin matrix of the various types of composites is independent of the type of filler phase, but dependent on monomer and the compositional factors of the resin matrix. It was also observed that volumetric contraction of these experimental composites appeared to depend not only on the type of resin, but also on the type of ACP.

Kim and coworkers [118] proposed bioactive bone cement (BBC), composed of natural bone powder (HAp), chitosan (CS) and commercially available PMMA-based bone cement. Investigators obtained three types of BBCs with different composition ratios: BBC I, BBC II and BBC III with 10 wt% of CS and 40, 50 and 60 wt% of HAp, respectively. Observation of the interfacial area between the host bone and the bone cement indicated that the BBC II composite has numerous pores that could be expected to afford space for bone ingrowth. However, after 4 weeks, the gaps between the host bone and the BBC II became narrower, and PMMA exhibited undesirable cleavage at the interfacial area; simultaneously, histological examinations of the interfaces at 4 weeks post-implantation demonstrated more new bone formations in the BBC II implant than in pure PMMA. In addition, the exothermic effects in the BBCs were considerably lower than that of pure PMMA.

7 Degradable Polymer/HAp (Nano)composites

For the last 40 years, the need for so-called "biodegradable" therapeutic material systems has received increasing attention in order to replace the use of biostable (or long-lasting) materials. This evolution is aimed at helping injured or diseased tissue to self-repair thanks to the outstanding healing ability of living systems. Among these domains, one can distinguish surgery with sutures, osteosynthesis devices (screws, plates, staples, etc.), pharmacology with drug delivery systems and tissue engineering [127]. Both synthetic and naturally derived biodegradable polymeric materials have been extensively investigated as biomaterials for bone tissue regeneration and reconstruction.

Again, for the reader's convenience, an overview of degradable polymeric materials (both synthetic and natural) that are used as polymer matrices for HAp is presented in Table 2.

Synthetic and natural degradable polymer/HAp (nano)composites are described in Sects. 7.1 and 7.2.

7.1 Synthetic Degradable Polymers

Synthetic biodegradable polymers are attractive candidate materials for short-term medical applications like sutures, drug delivery devices, orthopaedic fixation devices, wound dressings, temporary vascular grafts, stents, different types of tissue engineered grafts, etc. [128]. Hydrolytically degradable polymers (in most cases synthetic degradable polymers) are generally preferred as implants due to their minimal site-to-site and patient-to-patient variations compared to enzymatically (natural) degradable polymers [64]. Polymers prepared from glycolic acid and lactic acid have found a multitude of uses in the medical industry, beginning with biodegradable sutures, first approved in the 1960s [129]. Since that time, materials based on lactic and glycolic acid as well as other materials, including poly(dioxanone),

Table 2 Characteristics of degradable polymer/HAp (nano)composites

Polymer		Mode of degradation/ erosion	Polymer/HAp (nano)composite	Average size of HAp particles	HAp content	Young's modulus (GPa)	Tensile strength (MPa)	Comments	Ref.
Synthetic derived	Polylactide (PLA)	Hydrolytically degradable/bulk erosion	PLLA/HAp	0.3–1.3 mm	50, 80 wt%	4–10			[135, 136]
			PLLA/HAp	3.0 μm	20–50 wt%	2.0–2.4		Resorbable bone fixation devices (Super-Fixsorb and Super-Fixsorb-MX)	[143–145]
			PLA/HAp	NanoHAp	1–10 wt%			Anisotropy, open-cell	[148]
			PLA/β-TCP	MicroHAp					
			PLA/HAp	<35 μm	60 wt%	12		HAp powder was encapsulated in PLL-based microspheres	[3]
			PDLLA/HAp	2.0 μm	25–50 wt%			Shape memory properties	[150]
			PLA/PLA-*g*-HAp	100 × 20–40 nm	2–30 wt%				[149]
	Polyglycolide (PGA) and poly(lactide-*co*-glicolide) (PLGA)	Hydrolytically degradable/bulk erosion	PLGA/HAp	HAp fibre	10–40 wt%			Porogens: gelatin microspheres or NaCl salt particles	[154]

(continued)

Table 2 (continued)

Polymer	Mode of degradation/erosion	Polymer/HAp (nano)composite	Average size of HAp particles	HAp content	Young's modulus (GPa)	Tensile strength (MPa)	Comments	Ref.
		PLGA/HAp		10–30 wt%	2.3–4.6	4–10	Colloidal non-aqueous chemical precipitation technique at room temperature	[158]
		PLGA/HAp	100 nm	50 wt%			GF/PL method without the use of organic solvents	[159]
Poly(ε-caprolactone) (PCL)	Hydrolytically degradable/surface and bulk erosion	PCL/HAp	38–53 μm	10–30 wt%		20–45	Conventional blended in an extruder; grafted of PCL on the surface of HAp particles – ring opening polymerization	[165]
		Semi-IPN(l-PCL/net-PCL)/HAp	40–80 × 20–40 nm	0–15.8 wt%			Semi-IPN technique	[30]
		PCL/HAp	4 μm	13–32 vol%	21–28		Phase inversion and casting technique	[164]

(continued)

Table 2 (continued)

Polymer	Mode of degradation/ erosion	Polymer/HAp (nano)composite	Average size of HAp particles	HAp content	Young's modulus (GPa)	Tensile strength (MPa)	Comments	Ref.
		PCL/HAp		30 wt%			PCL-grafted nanophase HAp showed excellent colloidal stability in PCL solutions	[69]
Polyhydroxy-butyrate (PHB)	Hydrolytically degradable/ surface erosion	PHB/HAp	24.5 μm	10–30 vol%			Compounded, milled and compression moulded	[173]
		PHBHV/HAp PHBHV/TCP	24.5 μm 11.2 μm	10–30 vol%			Compounded, milled and compression moulded; PHBHV copolymer contained 12% mol of β-HV units	[169]
		PHBHV/HAp	20–30 nm	3–14 vol%			Solution casting method (chloroform)	[173]
Poly(propylene fumarate) (PPF)	Hydrolytically degradable/ bulk erosion	PPF-PA/HAp	125 μm	2.5–10 wt%			PPF/phloroglucinol triglycidyl methacrylate blend (25:75, 50:50 and 75:25 w/w); bone cement	[180]

(continued)

Table 2 (continued)

Polymer		Mode of degradation/ erosion	Polymer/HAp (nano)composite	Average size of HAp particles	HAp content	Young's modulus (GPa)	Tensile strength (MPa)	Comments	Ref.
	Polyphosphazenes	Hydrolytically degradable/ surface and bulk erosion (depend on R)	PN-EOB/HAp PN-POB/HAp					Precursor powder (TTCP and DCPA) at 5, 10 and 15 wt% in situ formation of HAp in the presence of PN, where HAp is chemically bonded to the polymer	[188–190]
	Polyanhydrides	Hydrolytically degradable/ surface erosion						Bone cement	[197]
	Polyurethanes (PUs)	Hydrolytically degradable							[198]
	Poly(amino acids)	Enzymatically degradable	PAsp/HAp					Obtained by direct synthesis of PAsp/HAp	[208]
	Polyacrylonitrile (PAN) and carbon fibres (CF)		CF/HAp		80 vol%	80–100		Hot pressed	[219, 220]

(continued)

Table 2 (continued)

Polymer		Mode of degradation/ erosion	Polymer/HAp (nano)composite	Average size of HAp particles	HAp content	Young's modulus (GPa)	Tensile strength (MPa)	Comments	Ref.
Natural derived	Collagen (Col)	Enzymatically degradable	Col/HAp	50 × 20 nm	80 wt%			Synthesized through a precipitation method with calcium hydroxide suspension and phosphoric acid solution containing Col	[225]
			Col/HAp	50–100 nm	80 wt%			Synthesized through simultaneous titration coprecipitation method	[226]
			Col/HAp		80 wt%			With glutaraldehyde; prepared by a simultaneous titration method	[228]
			Col/HAp	200–400 μm	50–70 wt%			HAp and Col were of bovine origin	[4]

(continued)

Table 2 (continued)

Polymer	Mode of degradation/ erosion	Polymer/HAp (nano)composite	Average size of HAp particles	HAp content	Young's modulus (GPa)	Tensile strength (MPa)	Comments	Ref.
		Col/HAp	Fibre diameter ~6 nm				HAp nanocrystals grew on the surface of these collagen fibrils with the *c*-axis of nanocrystals of HAp orienting along the longitudinal axis of the fibrils	[232]
Gelatin (Gel)	Enzymatically degradable	Gel/HAp	4 × 9 (high Gel conc.) to 30 × 70 nm (low Gel conc.)	5–40 wt% of Gel			Prepared using the biomimetic process; the development of HAp nanocrystals in an aqueous Gel solution was highly influenced by the concentration ratio of Gel to HAp	[237]
		Gel/HAp	100 nm	10–30 wt%			Fabricated by the co-precipitation of HAp within a Gel sol and further freeze-drying	[242]
		Gel/HAp	50–100 nm	1–4 wt% of Gel			Precipitation	[243]

(continued)

Table 2 (continued)

Polymer	Mode of degradation/ erosion	Polymer/HAp (nano)composite	Average size of HAp particles	HAp content	Young's modulus (GPa)	Tensile strength (MPa)	Comments	Ref.
		Gel/HAp	5 × 60–100 nm				Gel/HAp composite microspheres in a water-in-oil emulsion	[244]
Silk fibroin (SF)	Enzymatically degradable	SF/HAp	25 × 10 nm				Prepared at room temperature by a co-precipitation method	[252]
		SF/HAp			0.13	22	Multilayered films consisting of SF and HAp by alternating lamination	[253]
		SF/HAp	Nanosize	5 wt%			Poly[4-methacryloyl-oxyethyl trimellitate anhydride (4-META)]-grafted SF	[254]
Chitin and chitosan (CS)	Enzymatically degradable	Chitin/HAp		25–75 wt%			Processed into air- and freeze-dried materials	[261]
		CS/HAp	50 × 230 nm	20–100 wt%			Prepared by a co-precipitation method	[45]
		CS/Gel/HAp	<75 μm	30–70 wt%			Biomimetic scaffold by phase separation method	[262]

(continued)

Table 2 (continued)

Polymer	Mode of degradation/ erosion	Polymer/HAp (nano)composite	Average size of HAp particles	HAp content	Young's modulus (GPa)	Tensile strength (MPa)	Comments	Ref.
		CS/HAp	36–52 nm	5–10 wt% of CS			HAp composite bone paste with CS, using wet chemical method at low temperature	[267]
		CS/HAp	20 × 80 nm	30–80 wt%			Prepared by a co-precipitation method	[265]
		CS/Gel/HAp	17–52 nm				NanoHAp was formed in-situ on the surface of CS–Gel network films	[264]
Starch	Enzymatically degradable	SEVA/HAp SCA/HAp	<6.5 µm	30–50 wt%			Starch-based polymers and composites as potential biomaterials	[274]
		SEVA/HAp	5–6 µm				SCORIM	[275]
Cellulose (Cell)	Enzymatically degradable	Cell/HAp					HAp formation on cellulose cloth with the aid of citric acid	[277]

(continued)

Table 2 (continued)

Polymer	Mode of degradation/ erosion	Polymer/HAp (nano)composite	Average size of HAp particles	HAp content	Young's modulus (GPa)	Tensile strength (MPa)	Comments	Ref.
		BC/HAp	20–46 nm				Bacterial cellulose (BC); synthesized via a biological route by soaking phosphorylated and unphosphorylated BCs in SBF	[278]
Alginate (Alg)	Enzymatically degradable	Alg/Col/HAp	Nano size	50–68 wt%			Synthesized biomimetically	[283]
Hyaluronic acid (HYA)	Enzymatically degradable	Col/HYA/HAp	40–280 µm	90 wt%			Prepared by a co-precipitation method	[289]

poly(trimethylene carbonate) copolymers, and ε-caprolactone homo- and copolymers have been accepted for use as medical devices [53, 130, 131]. The functional groups susceptible to hydrolysis include ester, phosphazene, anhydride, carbonate, amide and urethane groups [132], so, logically, the most important synthetic biodegradable polymers that are currently used in biomedice are polyesters, polyphosphazenes, polyanhydrites, polyurethanes and poly(amino acids).

Synthetic biomaterials are generally biologically inert and have more predictable properties than natural polymers. For applications that need materials with a certain level of biological activity, strategies to incorporate biological motifs (e.g. HAp in the bone tissue regeneration area) onto synthetic polymers in the form of hybrid materials have been generally developed, but some novel procedures are currently being introduced [60].

7.1.1 Polyesters

Polylactide

Polylactide (PLA) is linear aliphatic thermoplastic and biodegradable polyester. High molecular weight PLA is generally produced by the ring-opening polymerization of the lactide monomer or bacterial fermentation of cornstarch or sugar beets. Commercially available PLA grades are copolymers of poly(L-lactide) with mesolactide or D-lactide. The amount of D enantiomer is known to affect the properties of PLA, such as glass transition temperature, melting temperature and degree of crystallinity. PLA has good mechanical properties, thermal plasticity and biocompatibility, and is readily fabricated, thus making it a promising polymer for various end-use applications in the biomedical field [132, 133].

The homopolymer of L-lactide poly(L-lactide) (PLLA), is a semicrystalline polymer. PLLA is about 37% crystalline with a melting point of 175–178°C and a glass transition temperature of 60–65°C. The degradation time of PLLA is much slower than that of poly(D,L-lactide) (PDLLA) requiring more than 2 years to be completely absorbed [134]. PLLA exhibits high tensile strength and low elongation and consequently has a high modulus, stiffness, chemical and impact resistance, good wear and friction characteristics, wide temperature range, and wide processing window. All these properties make PLLA more applicable than amorphous polymers for load-bearing applications, such as in orthopaedic fixation and sutures. PDLLA as an amorphous polymer having a random distribution of both isomeric forms of lactic acid, exhibits lower tensile strength and higher elongation and much more rapid degradation time (in comparison to PLLA) making it more attractive as a drug delivery system [53].

Enhancement of the bioactivity potential of PLA in the repair of bone fracture composites with HAp have been actively studied since the 1980s. During the last decade, investigations on preparation, mechanical properties, interface structure, biocompatibility and biodegradability of the PLA/HAp (nano)composites became even more intensive [3, 31, 135–150].

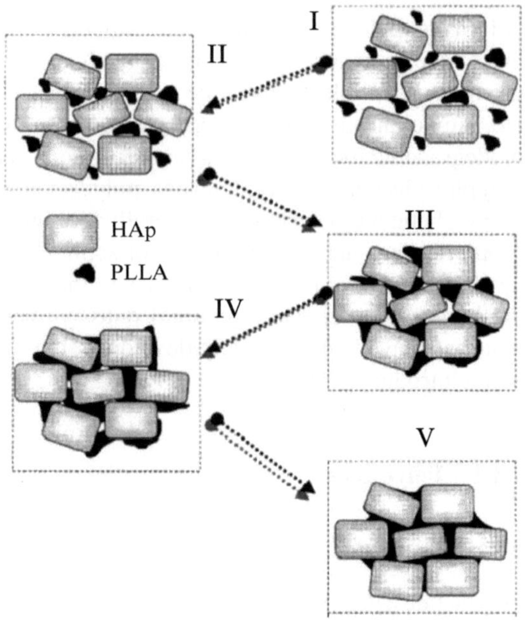

Fig. 10 PLLA/HAp liquid-phase sintering stages I–V. Reprinted from [135] with permission from Elsevier

Hence, Ignjatovic et al. [135] obtained PLLA/HAp composites by mixing completely dissolved PLLA with granules of HAp and compacting the system by cold and hot pressing at pressures of 49–490 MPa and temperatures of 20–184°C. The hot pressing temperature at which homogeneous systems were obtained (184°C) was close to the melting temperature of PLLA; a material with maximum porosity of 0.4%, density of 99.6% and compressive strength of 93 MPa was produced. The effect of cold and hot pressing temperature, pressure and time on the density, compressive strength and microstructure of the PLLA/HAp biomaterial was explained by a classical model of sintering in the presence of the liquid phase. Five stages of the process are shown schematically in Fig. 10.

Stage I corresponds to the system obtained by mixing HAp and solid PLLA, and cold pressing of this system gives a stage corresponding to stage II. Further treatment of the system by hot pressing via stages III and IV results in nonporous stage V. Addition of HAp granules into PLLA, previously dissolved in chloroform, followed by evaporation of the mixture, gives a system that mostly corresponds to the state found between stages II and III. For this reason, PLLA was first completely dissolved in chloroform and then mixed with HAp granules. However, the remaining chloroform might cause some adverse effects in the presence of tissue [135].

This group of researchers also presented results of in vivo investigations of the PLLA/HAp composites [136]. The composite that had been implanted into Balb/c Singen mice was removed after 1 and 3 weeks, and subjected to FTIR analysis. Formation of new functional groups and compounds during the bone repair process using PLLA/HAp biocomposites was observed The appearance of the absorption bands at ca. 1650 and 3420 cm^{-1} indicates generation of connective tissue of collagen, which was not present at the beginning of the experiment. Besides, bioresorption and disappearance of PLLA took place after implantation of PLLA-based/HAp biocomposites The same final products were generated using either PLLA/HAp biocomposite alone or in the mixture with autologous bone powder, which was evidence that fabricated materials can be successful substitutes for autologous bone powder [136, 137].

Jin and Gonsalves [138] applied poly(ε-caprolactone-*co*-vinyl phosphonic acid) (PCLVPA) and poly(ε-caprolactone-*co*-dimethylvinyl phosphoester) (PCLDMVPE) to the preparation of composites with PLLA and HAp. The PCLVPA was used as filler in PLLA films because it has pendant functional groups $P(O)(OH)_2$ that provide nucleation sites for the deposition of HAp in SBF. Generally, phosphorus-containing polymers are biocompatible and can provide nucleation sites for the deposition of inorganic minerals such as HAp. Phosphoprotein containing PO_3^{2-} side-chain groups in collagen compartments can localize a microenviroment containing free mineral ions and nucleate the mineral phase heterogeneously. Investigators found that HAp can grow on both the PCLVPA powder and PLLA–PCLVPA film in SBF solution; degradation occurs during biomineralization. The second material studied – a blend of PLLA and PCLDMVPE – has good hydrophilicity and was used to make a multilayered, porous structure containing HAp particles.

Ural and coworkers [139] synthesized two different elastomeric D,L-lactide and ε-caprolactone copolymers with different molecular weights (108,000 and 40,000) by ring-opening polymerization of the respective dimers, which were then loaded with different amounts of HAp in order to obtain bone-filling flexible materials for nonload-bearing applications. With regard to the mechanical properties, the percentage elongation decreased, while both the Young's modulus and the yield point (stress) increased with increased HAp loading. The copolymers were also degraded in the Ringer solution for 6 weeks: the molecular weight distribution became broader but, after incorporation of HAp, the degradation rate was significantly reduced. Due to the unique physical properties and suitable degradation rates, these HAp-containing bioactive composites are considered to be novel bone-filling materials; they have already been implanted in the bones of selected animals, and hard tissue regeneration without any side effects was observed [139].

Biodegradable lactide-based polymer/HAp composites have been prepared by Durucan and coworkers via a low-temperature chemical route [140] involving precomposite structures obtained by mixing α-$Ca_3(PO4)_2$ (α-TCP) with PLLA or poly(D,L-lactide-*co*-glycolide) (PLGA) copolymers. The final composite structure was achieved by in situ hydrolysis of α-TCP to $Ca_9(HPO_4)(PO_4)_5OH$ (cdHAp) either in solvent-casted or pressed precomposites. The presence of biodegradable polymers retards the formation of cdHAp in composite structures, whereby the transformation to cdHAp was more complete in precomposite structures made by solvent casting, compared with those made by pressing. The extent of transformation was also affected by the total porosity of the precomposites (Fig. 11).

Interestingly, the NaCl particles incorporated in the pressed precomposites as a pore generator improved the extent of transformation from 19 to 65% for PLGA (50:50) and from 26 to 81% for PLGA (85:15).

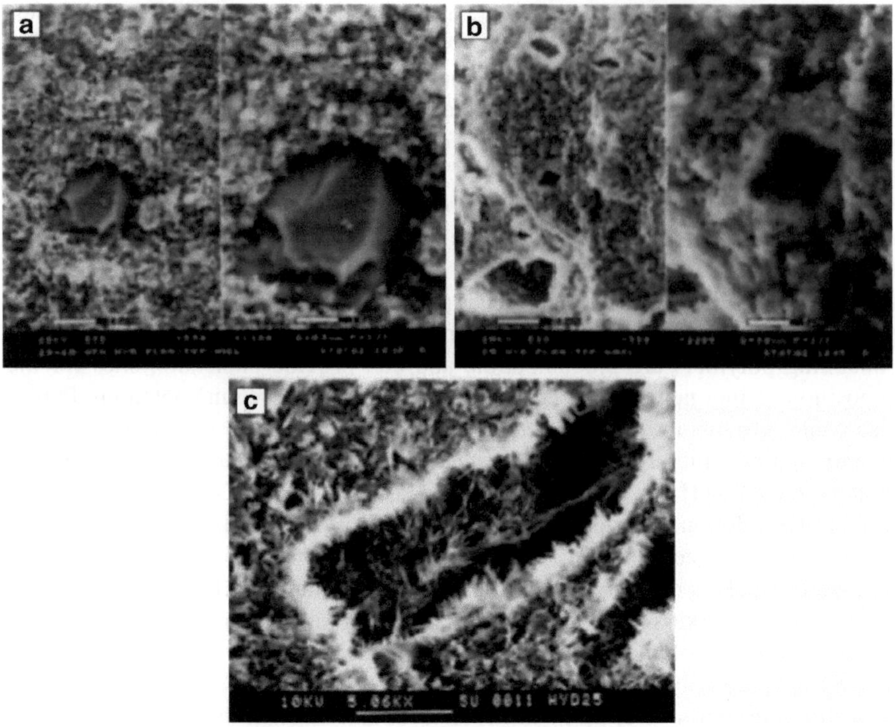

Fig. 11 SEM micrographs of 80:10:10 wt% β-TCP-PLGA (85:15)-NaCl composites. (**a**) Before hydrolysis (scale bars: 20 μm and 10 μm); (**b**) after hydrolysis at 56°C (scale bars: 20 μm and 10 μm); and (**c**) the morphology of cdHAp formed in the pore (scale bar: 5 μm). Reprinted from [140] with permission from Wiley InterScience

Deng et al. [31] obtained PLA/cdHAp nanocomposites by solvent casting. SEM observations confirmed close contact between the polymer matrix and the filled nanocrystals, and the homogeneous dispersion of nanocrystals in the polymer matrix at a microscopic level. The tensile modulus for the nanocomposites increased with cdHAp loading. Theoretical predictions of the modulus (by assuming that the nanocomposites behave as a short-fibre-filled system) based on the Halpin–Tsai equation show excellent agreement with the experimental results.

Biodegradation behaviour of PLLA/HAp composite rods for fracture repair was studied by Furukawa et al. [141]. Researchers used two kinds of composite materials: PLLA/uncalcined HAp (ucalHAp), which contained 30% by weight of ucalHAp, and PLLA/calcined HAp (calHAp) with 30 wt% of calHAp as a reinforcement. Composite rods were implanted in the subcutis and in the medullary cavities of rabbits. The specimen was then removed and examined at intervals between 2 and 52 weeks. The authors observed that the bending strength of the composites implanted in the subcutis was maintained at more than 200 MPa after 25 weeks and at 150 MPa after 52 weeks, whereby the molecular weight dropped to 45% of the initial value after 8 weeks and to approximately 10% after 52 weeks. PLLA/ucalHAp showed faster degradation rates than PLLA/calHAp composite, so the former has better resorption capability. Moreover, no inflammatory cells around the implants were found, which is a proof of excellent biocompatibility.

Further, long-term investigations on biocompatibility and biodegradation of these composites, revealed that four of 25 rabbits lived for more than 5 years after implantation, with one living for more than 7 years. For PLLA/ucalHAp, the shrinkage of the rods progressed further after 4 years, and some were almost completely resorbed [142].

In the work of Shikinami and Okuno [143], ucalHAp particles (20–50 wt%) were distributed in PLLA matrix to obtain composites by a forging process. The obtained composites generally exhibited the highest mechanical strength among this type of reinforced polymer composites with bioceramic fibres or particles known to date – the bending strength (S'') of about 270 MPa was far higher than that of cortical bone, and the modulus (E'') of 12 GPa was almost equivalent to that for cortical bone. Compared to PLLA devices only, these have ultrahigh initial strength (which is appropriately retained), bioresorbability, radio-opacity, and bioactivity such as osteoconductivity and bonebonding capability [143, 144].

The complete process of bioresorption and bone replacement of rods made of forged composites of PLLA/ucalHAp implanted in the femoral medullary cavities of rabbits was described by Shikinami et al. [145]. Researchers compared bioresorption, osteoconductive bioactivity and bone replacement in three implantation sites. In the first site, the end of the rod was located near the endosteum in the proximal medullary cavity and the rod was completely resorbed – unbound ucalHAp particles were detected in and around the endosteum 5–6 years after implantation. In the second case, the rod was located at the centre of the bone marrow space without contacting the endosteum – it showed significant shrinkage 4–5 years after implantation due to the release of almost all the PLLA, although a contracted cylindrical structure containing a few ucalHAp particles persisted even after 6 years. The rod in the third location was in direct contact with cancellous bone within the distal femoral condyle, and ucalHAp particles were almost completely replaced with bone after 5–6 years, as schematically displayed in Fig. 12.

For comparison, PLLA-only devices require longer periods (an additional 1–3 years) than composite devices for complete resorption [145].

Another approach was presented in the work of Russias and coworkers [3], who synthesized microspheres of HAp powder encapsulated in polylactide through an emulsion solvent evaporation method. The microspheres were then used as building blocks to produce dense and homogeneous composites by hot pressing. To examine the composites in 3D, synchrotron X-ray computed tomography investigations on both the microspheres and the subsequent hotpressed composites were performed Fig. 13.

The microstructure of the composites was quite homogeneous, although there were some instances of HAp agglomeration, as shown by white arrows in bottom right panel in Fig. 13.

Results of mechanical tests revealed that the modulus, strength and the fracture toughness of hot-pressed composites were comparable to those of human cortical

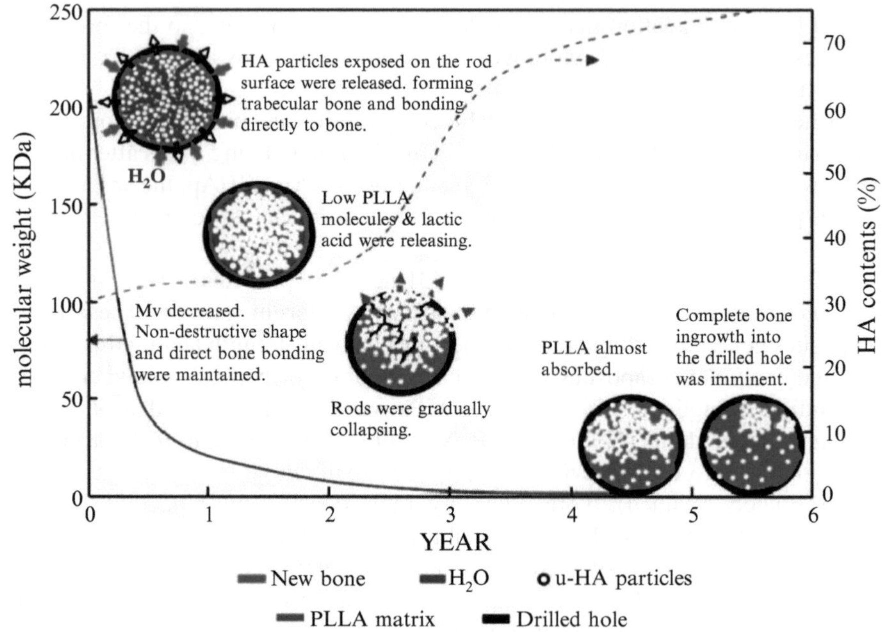

Fig. 12 Changes in morphology, Mv and uncalcined HAp (*u-HA*) content in rods in the distal femoral condyle. Reprinted from [145] with permission from Elsevier

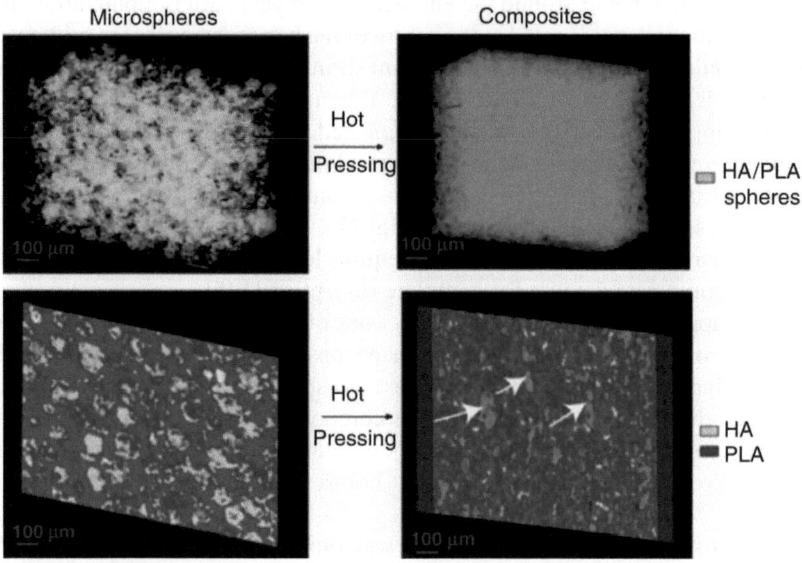

Fig. 13 3D reconstructed images (*top*) and 2D cross-sectional "slices" taken at the center of the images (*bottom*) obtained by synchrotron X-ray-computed tomography of the PLA/HAp microspheres (*left*) and of the composite obtained by hot pressing them at 190°C for 30 min under 90 MPa pressure (*right*). Reprinted from [3] with permission from Springer

bone. Moreover, the elastic modulus was relatively unaffected by in vitro degradation, and both the strength and the fracture toughness decreased with immersion in a simulated environment, presumably due to degradation of the polymer phase [3].

Hong and coworkers [146, 147] grafted PLLA directly onto the hydroxyl groups in the surface of nanoHAp particles by ring-opening polymerization of L-lactide in the presence of stannous octanoate [$Sn(Oct)_2$] as a catalyst. Investigators verified the existence of chemical linkage between PLLA and nanoHAp by ^{31}P NMR and FTIR analysis. The PLLA-*graft*-HAp particles could be more uniformly dispersed in chloroform than the non-grafted ones. When they are blended with PLLA via a solution process, the particles can be uniformly dispersed in the PLLA matrix and show improved adhesion with the PLLA matrix (Fig. 14).

Also, PLLA/*graft*-HAp materials had better mechanical properties than corresponding PLLA/HAp composites. At a *graft*-HAp (*g*-HAp) content of 4 wt%, the PLLA/*g*-HAp nanocomposite showed a maximum in tensile strength, bending strength and impact energy. The PLLA/*g*-HAp composites also demonstrated improved cell compatibility due to the good biocompatibility of the HAp nanoparticles and the more uniform distribution of the *g*-HAp nanoparticles on the film's surface [147].

Opencell composite foams comprising PLA and ceramic fillers, HAp or β-TCP were obtained by Mathieu and coworkers using supercritical CO_2 [148]. All the foaming conditions tested led to foams with structural parameters in the range of those of cancellous bone. This similarity is shown in Fig. 15, which compares three polymer foams and three different types of trabecular bones.

Compressive tests demonstrated anisotropy in mechanical behaviour, with an axial modulus up to 1.5 times greater than the transverse modulus. Composite scaffolds also showed viscoelastic behaviour with increased modulus for higher

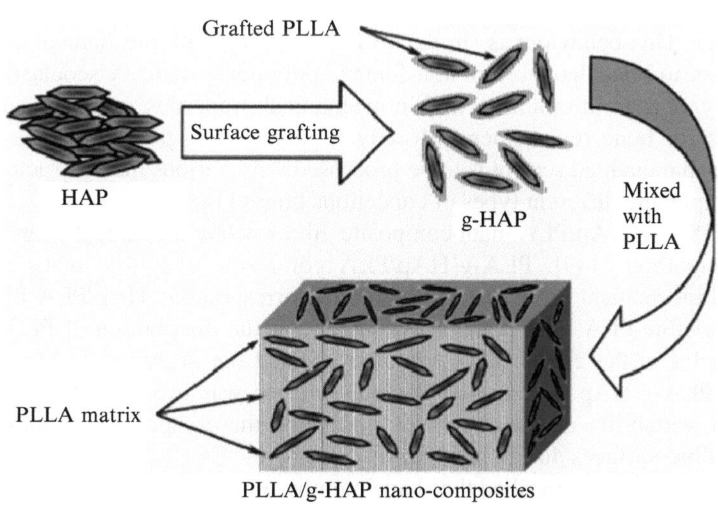

Fig. 14 Method for preparing PLLA/*graft*-HAp nanocomposites. Reprinted from [147] with permission from Elsevier

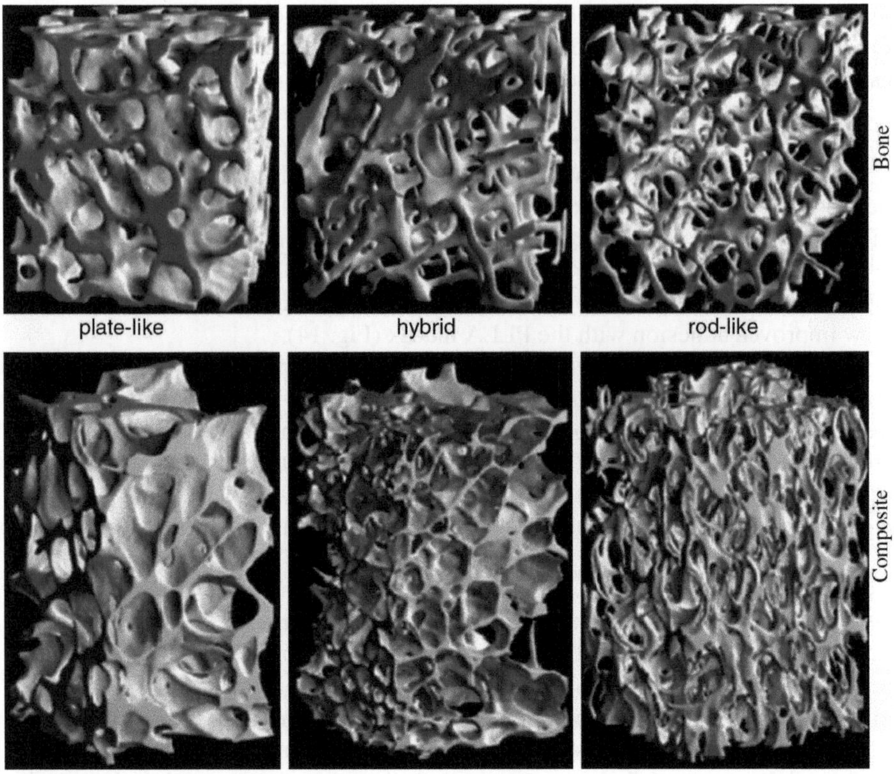

Fig. 15 Similarity of cancellous bone structures and PLA foam macrostructures. Reprinted from [148] with permission from Elsevier

strain rates. This behaviour is similar to that of bone, with mechanical characteristics closer to bone than current non-foamed polymer systems. Viscoelasticity and anisotropy, both in morphology and in mechanical properties, are promising characteristics for bone replacement. Moreover, gas foaming proved to be a flexible technique that enabled scaffolds to be processed with various macrostructures suitable for replacing different types of cancellous bones [148].

Also, PLA-*g*-HAp/PLA nanocomposite fibres were prepared by an electrospinning method [149]. PLA-*g*-HAp/PLA composite nanofibre mats exhibited improved mechanical properties compared to corresponding HAp/PLA fibre mats and the pristine PLA fibre mats. The content and the distribution of PLA-*g*-HAP nanoparticles in the composite fibres also affected the degradation rate: with increasing PLA-*g*-HAp content, the degradation rate was increased because of the enhanced wettability of the composite fibres and the escape of the nanoparticles from the fibre surfaces during incubation.

Zheng et al. [150] described the shape memory properties of PDLLA/HAp composites. The results show that the content and degree of dispersion of HAp particles in PDLLA matrix strongly influence the shape memory effect. The shape recovery ratios of the PDLLA/HAp composites were the biggest, and shape recovery time fastest, at a composite ratio between 2.0 and 2.5. The shape memory effect may be utilized for some implant applications that have to mimic the bone's shape in an exact way.

Gay and coworkers [151] studied dense PLLA/nanoHAp composites prepared by hot pressing. They proved that dense PLLA/HAp composites with superior mechanical performances, close to those of cortical bone, can be prepared by dispersing HAp nanoparticles into a PLLA matrix. Microscopic observation has shown that HAp particles remain uniformly dispersed at the nanoscale throughout the preparation steps. Mechanical tests have revealed that, on increasing the HA content, the strength and elastic modulus of the composite were significantly increased. At higher mineral contents, the rupture mechanism of the composite changes from ductile to brittle.

Polyglycolide and Poly(lactide-*co*-glycolide)

Polyglycolide (PGA) is the simplest linear aliphatic polyester, characterized by high degree of crystallinity (45–55%), high melting point (220–225°C) and a glass transition temperature of 35–40°C. PGA is not soluble in most organic solvents except for fluorinated organic solvents such as hexafluoroisopropanol. To reduce the stiffness of the resulting fibres glycolide has been copolymerized with other monomers [53, 152].

For instance, copolymers of glycolide with both L-lactide and D,L-lactide have been developed for orthopaedic and drug delivery applications [53]. However, foams produced from PLGA were found to have insufficient compressive strength to be used in any load-bearing application [153]. The use of PLGA-based composite materials to assist in the bone healing process, by either stabilizing the defect site or improving the osteoconductivity of existing biomaterials, has been also studied [154].

Hence, Thomson et al. [154] obtained 3D foam scaffolds of PLGA and HAp short fibres by solvent casting and compression moulding, followed by particulate leaching. By changing the weight fraction of the leachable component it was possible to produce composite foams with controlled porosities that have minimal compressive yield strength and can be seeded with cells. Low-porosity foams with enhanced osteoconductivity and compressive yield strength, which are however not suitable for cell seeding, can be also obtained by this method.

PLGA fibres with HAp particles differing in diameter have also been processed by solution spinning [155]. The authors found that HAp can be introduced into the fibres during the solution spinning process and/or deposited on the surface of the fibres. The presence of HAp particles in PLGA fibres accelerates deposition of calcium phosphate from SBF solution, as evidenced by analysis of SEM microphotographs. Moreover, X-ray analysis of the deposits indicated the presence of calcium, phosphorous, carbon and oxygen. The molar ratio of calcium to phosphorous was in the range 1.55–1.74, similar to that of stoichiometric HAp [155].

In a parallel study, the influence of bioactive additives, such as short carbon fibres (CF), HAp nanoparticles and bioglass, on the thermal and mechanical properties of PLGA was examined under in vitro conditions. The presence of bioactive particles affects the process of apatite growth on composite surface, in which a chemical bond between the implant and the bone tissue is formed. Despite the deterioration of mechanical properties after incubation under in vitro conditions, the PLGA/HAp composites still showed advantageous biological behaviour [156].

This team of investigators reported on the resorption process of PLGA/HAp nanocomposites implanted into rabbit submaxilla bones [157]. The resorption of composite occurred simultaneously with the formation of new bone tissue. The authors found that 3 weeks after implantation, the bone–implant interface becomes rough and polymer losses can be observed, with bone tissue in-growth. SEM microphotographs taken 12 weeks after implantation indicate the presence of fractal apatite grains growing towards the voids left after the resorbed polymer; the voids were completely filled by bone tissue a year after implantation.

PLGA/HAp composites were also produced by colloidal non-aqueous chemical precipitation at room temperature by Petricca and coworkers [158]. The shape of the scaffolds with homogeneous distribution of HAp particles in PLGA matrix was easily engineered, and the mechanical properties of the PLGA/HAp composites indicated that the materials are suitable as bone substitution scaffolds. The decreased tensile strength was attributed to the presence of micropores within the scaffold, which are created upon removal of the ammonium nitrate, a by-product of the synthesis. The novelty of the process is the ability to incorporate up to 30 wt% HAp within the polymer matrix, thereby maximizing the material osteoconductivity, as evidenced by osteoblast-like cell adhesion experiments.

Gas foaming and particulate leaching (GF/PL) was applied for fabrication of PLGA/HAp scaffolds, too, according to the scheme depicted in Fig. 16.

Since the highly crystalline HAp degrades in vivo over a long period of time, incompletely decomposed residual HAp may hinder or slow complete bone healing. Therefore, to reduce the total amount of HAp, nanosized HAp particles were used to fabricate the composite scaffolds. Furthermore, nanosized HAp particles show improved bioactivity and osteointegration when implanted into the bone defect site, compared to the microsized HAp particles, and show better protein adsorption and cell adhesion. Investigators observed that PLGA/HAp composite scaffolds, fabricated by the GF/PL method, show enhanced hydrophilicity and osteoconductivity compared with the solvent casting and particulate leaching (SC/PL) scaffolds, and this enhancement was most probably due to a higher extent of exposure of HAp particles to the scaffold's surface. The biodegradable polymer/bioceramic composite scaffolds fabricated by the GF/PL method could thus enhance bone regeneration efficacy for the treatment of bone defects in comparison with conventional composite scaffolds [159].

Zhang et al. [160] prepared a 3D porous scaffold with the composite of PLGA and PLLA-grafted nanoHAp using a SC/PL method. They revealed that the nanoparticles of *g*-HAp distributed on the surface of PLGA matrix more uniformly than that of ungrafted HAP. Moreover, the intramuscular implant study showed that

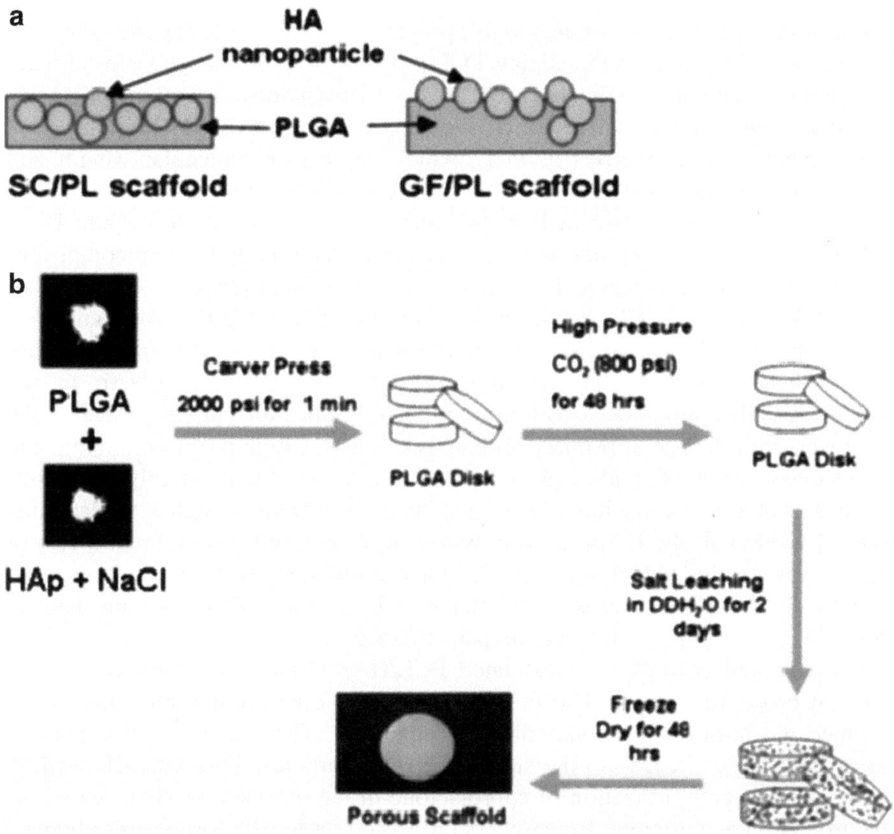

Fig. 16 (**a**) Exposure of HAp nanoparticles to the surface of scaffolds fabricated by the GF/PL and SC/PL methods. (**b**) Fabrication of PLGA/HAp composite scaffolds with the GF/PL method. Adapted from [159]

there were two outcomes for the nanocomposite of *g*-HAp/PLGA: mineralized or completely biodegraded. This may depend on the content of HAp in the polymer matrix and on the balance of biodegradation, mineralization and tissue regeneration. As the bone replacement, the scaffold of *g*-HAp/PLGA exhibited rapid and strong osteoconductivity for repairing critical radius defects.

Poly(ε-caprolactone)

Poly(εcaprolactone) (PCL) is linear polyester manufactured by ring-opening polymerization of ε-caprolactone. It is a semicrystalline polymer with a degree of crystallinity of ca. 50% and rather low glass transition temperature (−60°C) and melting point (65°C). PCL exhibits high elongation at break and a low elastic modulus. Its physical properties and commercial availability make it attractive not only as a substitute material for nondegradable polymers for commodity applications, but also as a specific polymer in medicine. PCL properties can be changed substantially by blending it with other polymers [161, 162] or by radiation crosslinking processes resulting in enhanced properties for a wide range of applications [133].

The mechanical properties of PCL, which depend on molecular weight and preparation method, usually do not allow its use as scaffold for hard tissue replacement. Incorporation of HAp in PCL has improved (in comparison to pure PCL) both the mechanical properties and osteoblast growth on the PCL/HAp composites prepared by phase inversion and solution casting techniques [163].

Scaffolds made of PCL and HAp were a topic of investigations of Causa and coworkers [164]. They found that the mechanical properties of the composites are close to those of human bone only after the addition of 20 vol% of HAp. In particular, the elastic modulus is within the range of values for human cortical bone. Moreover, with the use of primary human osteoblasts, a high proliferation rate and a moderate increase of alkaline phosphatase activity were found, mainly on the surface of PCL-based composites with 13 and 20 vol% of HAp, though at the last time point (4 weeks) all the HAp-added polymers were covered by confluent layers of cells. It was concluded that the structure of a scaffold along with its surface physicochemical characteristics affect cell behaviour, but, on the other hand, mechanical properties are also crucial for implant performance.

Azevedo and coworkers investigated PCL/HAp composites prepared by two different procedures [165]. The first consisted of a conventional blending of the polymer and reinforcement material in an extruder. The second method encompassed grafting of PCL onto the surface of HAp particles. This was achieved by a ring-opening polymerization of caprolactone in the presence of HAp, where its OH groups act as initiators. Researchers observed that for the composites obtained by both methods an increase in the modulus occurred for higher amounts of filler. Degradation results were in line with this observation, as the grafted composites uptake less water and degrade at a slower rate. The authors suggested that this method for surface modification of the HAp particles is a promising route for obtaining composites with enhanced properties to be used in orthopaedics.

In work of Lee et al. [69], the effect of surface-modified HAp nanocrystals on the biocompatibility of PCL-based materials was described. These materials were obtained by surface grafting of PCL on HAp modified by ethylene glycol (Fig. 17).

The authors observed that PCL-grafted HAp nanophase showed excellent colloidal stability in PCL solutions, which ensures the nanolevel distribution and the large specific surface area of HAp in the nanocomposites. In vitro biological evaluation showed that the existence of modified HAp in nanocomposites offered favourable environments for protein adsorption and cell adhesion and proliferation.

Hao et al. [30] obtained a semi-interpenetrating polymer network (semi-IPN) based on linear PCL (l-PCL) and the PCL network (net-PCL) in which HAp nanocrystals were homogeneously dispersed. The preparation process of semi-IPN (l-PCL/net-PCL)/HAp is shown in Fig. 18.

In semi-IPN matrix, phase separation occurs The morphology of net-PCL is featured by intracrosslinked microdomains that further interconnect with each other to

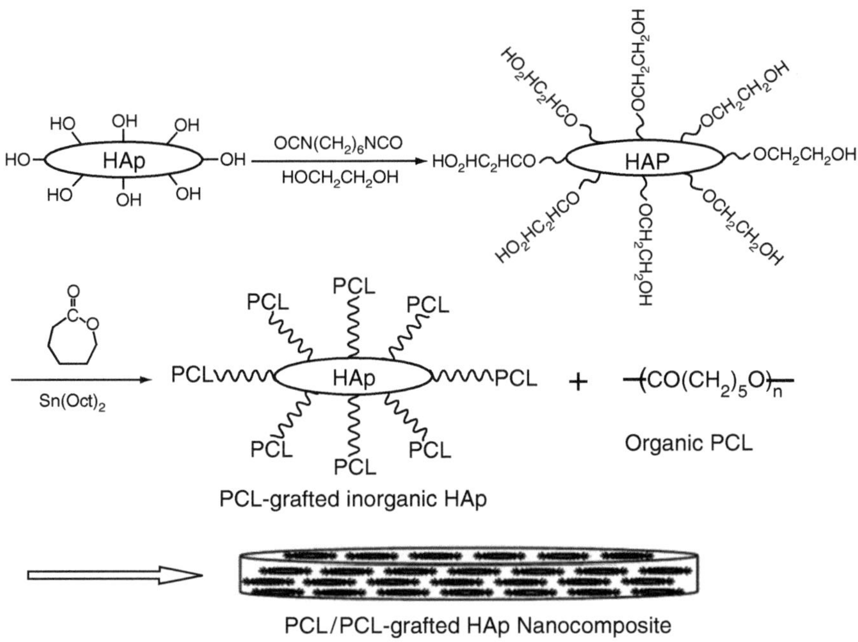

Fig. 17 Surface grafting of PCL onto nanophase HAp. Reprinted from [69] with permission from Elsevier

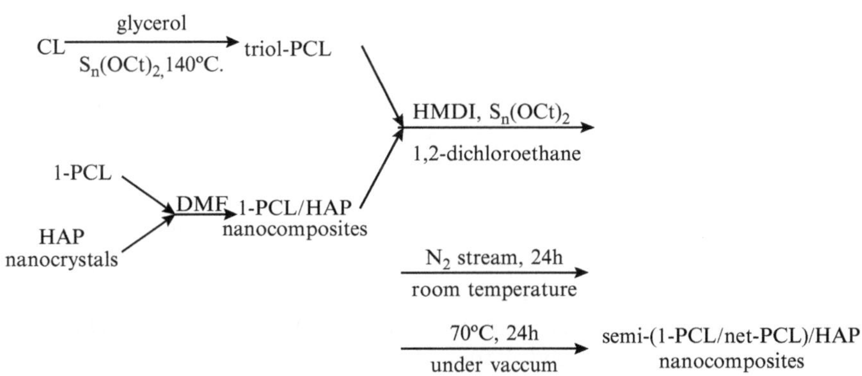

Fig. 18 Process for preparation of semi-IPN (l-PCL/net-PCL)/HAp nanocomposites. Reprinted from [30] with permission from Elsevier

form the network, whereby l-PCL may crystallize separately. The net-PCL component is totally amorphous at room temperature for the nanocomposites containing HAp up to 12.3 wt%. Further, the crystallinity of l-PCL is greatly decreased due to the presence of net-PCL as compared with that for pure l-PCL. The incorporation of l-PCL into the amorphous net-PCL significantly improved the mechanical properties of the system. A great improvement in mechanical properties is observed for the nanocomposites if the HAp content is increased to 15.8 wt%, which is associated with a transition of the net-PCL component from an amorphous to semicrystalline arrangement at this composition range.

Polyhydroxybutyrate and Poly(hydroxybutyrate-*co*-hydroxyvalerate)

Polyhydroxybutyrate (PHB) is a naturally occurring biodegradable β-hydroxyacid produced by various bacteria (enzymatic synthesis) or by several chemical synthetic routes. PHB is a semi-crystalline isotactic polymer that undergoes surface erosion by hydrolytic cleavage of the ester bonds and has a melting temperature in the range of 160–180°C [166–168]. PHB usefulness is limited by its brittleness; however, the addition of polyhydroxyvalerate (PHV) to the PHB can improve the ductility and processability of the (co)polymer. The copolymers of PHB and 3-HV (PHBHV) have similar semicrystalline properties as PHB, but the melting temperature is lower depending on the HV content [169, 170]. The copolymers can be processed using conventional manufacturing technologies, such as extrusion and injection moulding, or can be cast from solution [60, 167, 171, 172].

Many efforts have been made to improve the mechanical properties of PHB and PHBHV. The combination of osteoconductive properties of HAp with the processability of biodegradable polymer can overcome limitations induced by the brittleness and can widen the application potential in bone tissue engineering.

Hence, Wang and coworkers [173] obtained PHB/HAp composites by using a standardized procedure that consisted of compounding, milling and compression moulding. A biologically active apatite layer formed within a short period on PHB/HAp composite after its immersion in SBF, demonstrating high in vitro bioactivity. Furthermore, the bioactivity and mechanical properties of the composite could be tailored by varying the HAp volume percentage in the composite. The storage modulus of the composite increased initially with immersion time in SBF (due to the formation of the apatite layer on composite surface) and decreased after prolonged immersion in SBF, displaying degradation of the composite in a simulated body environment. Compounded and compression moulded PHBHV/HAp and PHBHV/β-TCP composites containing up to 30 vol% of the bioceramics were characterized using various techniques, including SEM, TG, DSC and DMA. The results demonstrated that intended compositions of blends had achieved and that bioceramic particles were well distributed in the polymer matrix. The thermal stability of PHBHV was slightly reduced by the incorporation of bioceramics, and the melting temperature was slightly affected. The storage modulus and loss modulus of the composites increased with the increase in HAp or β-TCP content. Composites containing the highest percentage of bioceramics exhibited the highest stiffness. Preliminary in vitro study indicated an enhanced ability of the composites to induce the formation of bone-like apatite on their surfaces [171].

PHBHV/HAp nanocomposites have been successfully fabricated by Chen and coworkers by using solution casting with the aid of strong ultrasonication [174]. The reinforcing effect of nanoHAp on the polymer matrix was confirmed by DMA data, from which a noticeable increase in the storage modulus, due to the addition of nanoHAp, was detected. The in vitro investigations indicated that the nanocomposites have an improved bioactivity over the conventional microstructured materials.

More recently, Noohom et al. [175] blended nanoHAp particles, stabilized using poly(acrylic acid), with poly(hydroxybutyrate-*co*-hydroxyvalerate) (PHBHV) using a precipitation/gelation method to produce PHBHV/HAp nanocomposites. The authors found that preparative methods that lead to an increase in the dispersion of the particles caused an increase in the compressive modulus as a function of particle loading, which was due primarily to a higher surface area of the filler particles. In addition, the higher degree of dispersion leads to a small increase in the overall crystalline content of the composites, which might also contribute to the increased stiffness due to the formation of a crystalline layer close to the particle interface. The obtained PHBHV/HAp nanocomposites possessed higher moduli and compressive strengths compared to those of cancellous bone, but lower than those of cortical bone [175].

Poly(propylene fumarate)

Poly(propylene fumarate) (PPF) is an unsaturated linear polyester that can be cross-linked through the fumarate double bond. Several routes, including the transesterification of fumaric diester, can be used to synthesize linear PPF. It is known to undergo bulk erosion via hydrolysis of the ester bonds and the degradation time depends on several parameters, such as molecular weight, type of cross-linker, and crosslinking density. The degradation products are propylene glycol, poly(acrylic acid-*co*-fumaric acid) and fumaric acid, a substance that occurs naturally as a part of the Kreb's cycle [65].

Several attempts have been made to improve the mechanical properties of PPF for orthopaedic applications by cross-linking or by developing composites with ceramic materials. The cross-linked matrices demonstrated a compressive strength in the order of 1–12 MPa depending on the composition and conditions of polymerization [60, 176]. In order to improve these properties for use in orthopaedics, PPF is often combined with particles of ceramic materials such as β-TCP, calcium carbonate or calcium phosphate [177–180]. These composite materials exhibit compressive strength of 2–30 MPa, which is appropriate for replacement of cancellous bone. These cross-linked polymer matrices also supported good cell viability and could function as a growth factor delivery systems, making them promising candidates for bone tissue engineering applications [60, 65].

Jayabalan et al. [180] studied PPF/phloroglucinol triglycidyl methacrylate/HAp bone cement for binding bovine rib bones. The tensile strength of this adhesive bond was found to be 1.11 kPa. Thermal studies suggest the onset of cross-linking reaction in the cured blend if the blend is heated. The in vitro biodegradation studies revealed possible association of calcium ions with negatively charged units of degrading polymer chain, resulting in slow-down of the degradation process. The obtained results make it possible to consider the potential use of PPF/phloroglucinol triglycidyl methacrylate/HAp as a partially degradable polymeric cement for orthopaedic applications.

Poly(ortho esters)

Poly(ortho esters) (POE) have been investigated since 1970 and four polymer families (POE I – POE IV) have been developed. Among these polymers, POE IV has significant potential for producing useful, commercially relevant bioerodable drug delivery products and for other biomedical applications, e.g., for bone tissue regeneration. The success of POE IV is largely based on the versatility of the synthesis which allows excellent control over erosion rates by selecting the proper ratios of diol to latent acid diol, and excellent control of mechanical properties by appropriate selection of diols to use in the synthesis. Poly(ortho esters) are thermoplastic materials that can be easily fabricated by extrusion injection moulding or compression moulding [181, 182].

The advantage of these polymers for bone tissue regeneration is that a new tissue grows into the space occupied by the artificial scaffolding and only the surface of the polymer scaffold degrades, leaving the bulk of the material with its original mechanical integrity. Solheim and coworkers investigated the effect of bone regeneration of PDLLA and POE IV by implanting samples containing polymer and demineralized bone particles (HAp-containing materials) into the abdominal muscle of rats. They found that POE showed no inflammation and had littletono effect on bone formation, whereas PDLLA provoked a chronic inflammatory response and inhibited bone formation [1, 183].

7.1.2 Polyphosphazenes

Polyphosphazenes are hybrid polymers with a backbone of alternating phosphorus and nitrogen atoms containing two organic side groups attached to each phosphorus atom. The side groups attached to the phosphazene backbone have a profound effect on the chemical, physical and mechanical properties of the resulting polymer; hence, specific polymers can be designed and synthesized for particular applications [60, 184, 185]. Biodegradable polyphosphazenes with highly controlled degradation rates can be synthesized by incorporating side groups such as imidazoyl, amino-acid esters, glycosyl-, glyceryl-, lactic- or glycolic-acid esters. Furthermore, polyphosphazenes are excellent materials for orthopaedic application due the high osteocompatibility of various biodegradable polyphosphazenes as evidenced from the in vitro and in vivo evaluations [186].

Polyphosphazenes can be also blended with ceramic materials that offer a range of interesting properties. TenHuisen et al. [187] studied the formation of composites based on poly[bis(sodium carboxylatophenoxy)phosphazene] and cdHAp at a low temperature. They found that the Ca^{2+} ions that were released during dissolution of the cdHAp precursors exchanged with the monovalent Na^{+} ions in the polymer, leading to a crosslinking and precipitation of the latter. The formation of cdHAp and the crosslinking of the polymer occurred concurrently, which led to the formation of a composite with the crosslinked polymer well incorporated into the composite material. However, the Na^{+} ions liberated caused the elevation of pH to cytotoxic values [187].

Greish et al. [188] described the formation of biodegradable polyphosphazene/HAp composites produced via an acid–base reaction of TTCP and anhydrous dicalcium phosphate in the presence of polyphosphazenes bearing side groups containing an alkyl ester. The polyphosphazenes used were poly(ethyl oxybenzoate)phosphazene (PN-EOB) and poly(propyl oxybenzoate) phosphazene (PN-POB). The results demonstrated that in the presence of PN-EOB or PN-POB, the generation of alkaline conditions promotes the partial hydrolysis of the alkyl ester side groups resulting in the formation of free carboxylate groups on the polymer. Moreover, calcium ions in solution cross-link these carboxylate groups, leading to the formation of a calcium crosslinked polyphosphazene surface layer, which in turn nucleates the deposition of HAp on the surface. The in situ formation, as well as the biomimetic deposition of HAp within the surrounding polymers, closely resembles the mineralization of collagen in vivo [188].

Formation of stoichiometric HAp (stHAp) via an acid-base reaction between TTCP and acidic calcium phosphate (e.g., $CaHPO_4$, DCPA) in the presence of poly[bis(sodium carboxylatophenoxy) phosphazene (Na-PCPP) and poly[bis(potassium carboxylatophenoxy) phosphazene (K-PCPP) at 37.4 and 50°C was also investigated. Inhibition of stHAp formation depends on the proportion of polymer used and on temperature. Moreover, to minimize the inhibition of formation of HAp, it was suggested that lower molecular weight polymers be used [189].

In subsequent work, Greish and coworkers [190] investigated the formation of SHAp in the presence of poly[bis(carboxylato phenoxy) phosphazene] (acid-PCPP), its ethyl ester (ethyl-PCPP) and propyl ester (propyl-PCPP). In the first step, the authors prepared composites made of acid-PCPP and HAp at physiological temperature. Researchers found that the rate of composite formation increased with the proportion of phosphazene. Furthermore, the higher concentrations of calcium and phosphorous ions found in solution during composite formation, compared with polymer-free HAp formation, indicate the effect of the acid-PCPP in enhancing the liberation of calcium and phosphorous ions to solution; the presence of acid-PCPP promoted reductions in pH during HAp formation. In the presence of 15 wt% acid-PCPP, pH excursions to cytotoxic values were avoided. This effect, coupled with accelerated formation of HAp in the presence of acid-PCPP, is advantageous in terms of in vivo formation of bone analogues [190].

7.1.3 Polyanhydrides

Polyanhydrides containing the most reactive functional group available for degradation on the base of passive hydrolysis, are synthesized via melt condensation of diacids/diacid esters, ring opening polymerization of anhydrides, interfacial condensation, dehydrochlorination of diacids and diacid chlorides or by the reaction of diacyl chlorides with coupling agents such as phosgene or diphosgene [62, 191]. Polyanhydrides are generally classified as surface-eroding polymers because they undergo a linear mass loss during erosion. The mechanical properties of these polymers were found to be less than optimal for loadbearing applications, such as for orthopaedic implants. The search for high strength polyanhydrides with surfaceeroding properties has led to the development of poly(anhydride-*co*-imides). For instance, Attawia and coworkers have investigated the mechanical performance and biocompatibility of a wide range of poly(anhydride-*co*-imides), e.g. poly[pyromellitylimidoalanine-*co*-1,6-bis(p-carboxyphenoxy) hexane], as scaffolds for bone tissue engineering [192].

In another approach, photo-crosslinked polyanhydrides in the form of a 3D network for high mechanical strength and slow degradation have been obtained [91–94]. These polyanhydrides are useful as degradable orthopaedic fixation devices, for example pins and screws; resorbable fillers for bone augmentation and regeneration; bone cements with HAp, etc. A compressive strength similar to the lower range of cancellous bone (30–40 MPa) has been reported for this class of polymers [197].

7.1.4 Polyurethanes

Traditionally, polyurethanes (PUs) have been widely used in the biomedical field for their excellent mechanical properties and relatively good blood compatibility. PUs can offer many advantages in the design of injectable and biodegradable polymer compositions [198] because they have an established record of biocompatibility [199, 200], the ability to be functionalized to improve cell growth and proliferation, and controllable degradation kinetics [201]. Additionally, the mechanical properties of PUs can be tailored for uses as soft [202] and hard tissue [203, 204] biomaterials. Few studies have been devoted to developing PU injectable materials suitable for orthopaedic applications. These materials would require the additional property of having good mechanical properties and controlled degradability [60], so ceramic particulates should be the first choice as reinforcement.

Recently, Bonzani and coworkers [198] have synthesized and characterized a novel injectable and in situ curable PU-based material for tissue-engineered implants using pentaerythritol-based prepolymers. This polymer matrix displayed mechanical strength and elasticity superior to many injectable bone cements and grafts. Cytotoxicity tests using primary human osteoblasts revealed positive cell viability and increased proliferation over a period of 7 days in culture. Furthermore, the incorporation of β-TCP was shown to improve mechanical properties, surface wettability, and cell viability and proliferation compared to the other sample types. Based on preliminary mechanical analysis and cytotoxicity results, these injectable PUs may have a number or potential orthopaedic applications ranging from bone glues to scaffolds for bone regeneration [198].

7.1.5 Poly(amino acids)

Poly(L-glutamic acid) (L-PGA) and poly(aspartic acid) (PAsp) are synthetic poly(amino acids) that undergo enzymatic biodegradation. Several synthetic routes have been investigated for the synthesis of L-PGA; however, the triethylamine-initiated polymerization of the N-carboxyanhydride of *g*-benzyl-L-glutamate is the most widely used route [205]. A biosynthetic method for forming monodisperse L-PGA has been also applied, as well as a microwave-assisted procedure for synthesis of PAsp [206].

It was reported that the acidic macromolecules present in the ECM of biologically calcified tissues play a key role in the control of the polymorphism, orientation and growth of the inorganic crystals [207]. They have therefore been used in vitro to prepare polymer/HAp composites with reduced crystal dimensions for hard tissue replacement [208]. A high aspartic and glutamic amino acid residue content is considered to be an important requirement of the matrix proteins involved in the nucleation of calcium phosphates [209]. Glutamate seems to be involved in bone cell signalling and to play a role in mechanisms associated with intercellular communication [210] and with maintenance of homeostasis. Aspartic acid also seems to induce osteoblast differentiation and to increase extracellular mineralization. Furthermore, in vitro studies indicate that aspartic acid can act as a nucleating agent for apatite and influence its further growth [209, 211].

Boanini et al. [211] obtained Asp/HAp and L-glutamic acid/HAp composite nanocrystals, with an amino acid content up to 7.8 and 4.3 wt%, respectively, prepared by direct synthesis in aqueous medium. They found that incorporation of the amino acids (which occurs through a specific interaction with the HAp structure) affects the degree of crystallinity of the apatitic phase, which decreases as the amino acid content increases. The presence of amino acids controls the crystal dimensions and favours osteoblast proliferation, activation of their metabolism and differentiation, which is of high importance for potential biomedical applications. Direct synthesis of PAsp/HAp nanocrystals was performed and it was found that PAsp is quantitatively incorporated into HAp crystals, provoking a reduction of the coherent length of crystalline domains, as evidenced by WAXS and TEM observations [208].

7.1.6 Polyacrylonitryle and Carbon Fibres

Carbon fibres (CF) are obtained through carbonization of polyacrylonitrile (PAN). They have excellent mechanical properties and have been applied in HAp-containing biocomposites for a number of years in the form of short, continuous and textile CF materials [212–227]. The incorporation of CF as the modifying phase for a bioresorbable polymer matrix increases the strength of the composite, and the fibres uncovered after polymer resorption can serve as the scaffold for the treated bone [218]. Slosarczyk et al. [219] obtained composites from HAp powder and CF by hot pressing. The properties of CF/HAp composites depend significantly on the characteristics of the initial fibres and especially on the type of functional groups existing on their surface; fibres containing hydroxyl groups on the surface were preferred. The best homogeneity of composites throughout the entire volume, the best sintering grade and the highest strength properties were observed in the case of composites produced from CF coated with calcium phosphates, characterized by the basic character of the surface.

The same method was applied by Dorner-Reisel and coworkers to obtain CF/HAp composites [220]. Reinforcement of calcium HAp with 20 vol% of short CF positively improves the resistance against microabrasion. Additional improvements of the wear behaviour can be reached if the microstructure of the CF/HAp composite material is optimized. The authors concluded that the critical factor seems to be the mismatch between the thermal expansion of matrix and reinforcement.

Piekarczyk-Rajzer and Blazewicz [221] investigated in vitro behaviour of CF/HAp composite in physiological fluids and it was found that modification of the CF phase composition at the precursor stage could be a useful way to control the bioactive properties of CFs. Janowska and coworkers [222] studied the thermal properties of PAN-based fibre precursor containing nanoparticles of additives such as SiO_2, HAp and montmorillonite. The thermal stability of nanoadditive-containing PAN fibres depended both on the type of the incorporated nanoadditive and on the porous structure produced during fibre formation. The incorporation of nanoadditive into the fibre-forming polymer of precursor PAN fibres caused changes in the temperatures of characteristic thermal transitions, which highlights the necessity of proper selection of the carbonization parameters.

7.2 *Natural Degradable Polymers*

7.2.1 Proteins

Proteins, as amino acid polymers arranged in a 3D folded structure, are the major structural components of many tissues. They are preferred biomaterials for scaffolds for tissue engineering, and for use as sutures, haemostatic agents and drug delivery vehicles [223].

Collagen

In the search for a new generation of bone substitutes that are bioactive and biodegradable, composites consisting of collagen and calcium phosphate minerals have received much attention because they mimic the basic composition of bone (see Sect. 2, Bone structure) [224].

Biomimetic configurational arrays of HAp nanocrystals on collagen (Col) and chondroitin sulphate (ChS) were comparatively studied by Rhee et al. [225]. The nanocomposites of Col/HAp, ChS/HAp, and Col/ChS/HAp were synthesized through a precipitation method with calcium hydroxide suspension and phosphoric acid solution containing Col, ChS, and their mixture, respectively. The (002) diffraction of the synthesized HAp crystals on a Col fibre showed around 60° arching angle, whereas that on a ChS fibre showed just around 10° (Fig. 19).

Exactly the same configurational arrays of HAp crystals were obtained on the mixture of Col and ChS fibres. The results may be applicable for development of

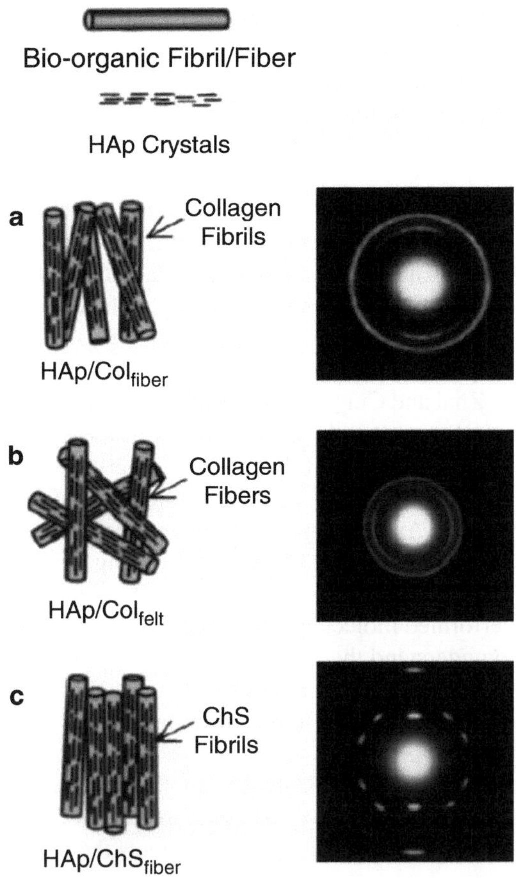

Fig. 19 Different electron diffraction patterns of HAp crystals caused by the different macromolecular configurations of bio-organics: (**a**) Col/HAp fibre; (**b**) Col/HAp felt; and (**c**) ChS/HAp fibre. Reprinted from [225] with permission from Elsevier

a bone substitute that mimics the peculiar configurational arrays of HAp crystals found in bone and its detailed organic composition.

Kikuchi et al. [226] synthesized a Col/HAp composite by a simultaneous titration coprecipitation method using $Ca(OH)_2$, H_3PO_4 and porcine atelocollagen as starting materials. The composite obtained showed a self-organized nanostructure, similar to bone, assembled by the chemical interaction between Col and HAp. The composite was consolidated by a cold isostatic pressure of 200 MPa and indicated a quarter of the mechanical strength of bone and the same biological properties as grafted bone. The material was then resorbed by phagocytosis of osteoclast-like cells and conducted osteoblasts to form new bone in the surrounding area. The short-range cross-linkage with glutaraldehyde was successfully induced to the Col/HAp composites. The linkage between the self-organized units restricts water intrusion into the composite fibrils at low glutaraldehyde concentrations. By increasing the glutaraldehyde amount, interfibril cross-linkage with long glutaraldehyde chains increases and maintains large amounts of water even after applying pressure. The cross-linkage of the composite enhanced its mechanical strength by both inter- and intrafibril linkages. In parallel to that, the in vivo resorption rate was reduced without any toxic reactions, including inflammation [227–231].

Biomaterials composed of ColI and sintered (at 1100°C) or powdered HAp, both of bovine origin, were obtained by Rodrigues et al. [4] for osteoconductive and osteoinductive scaffolds. ColI:HAp proportions were 1:2.6 and 1:1 (wet weight), and particles sizes varied from 200 to 400 μm. The authors suggested that the biomaterials analysed and characterized in this work had some properties and characteristics reported by the literature as optimum for enhancing the generation of a new bone tissue. These characteristics included the diameter of particles, presence of Col, optimal macroporosity caused by collagen–particle interaction, and the presence of weakly bound carbonate groups in the HAp that can potentially generate carbonate–HAp on the particle surface.

Zhai and Cui [232] reported that recombinant human-like ColI, an acidic protein, can direct growth of HAp nanocrystals in vitro in the form of self-assembly of nanofibrils of mineralized Col (Fig. 20) that resembles the ECM.

The mineralized Col fibrils aligned parallel to each other to form mineralized Col fibres. HAp nanocrystals grew on the surface of these Col structures with the *c*-axis of nanocrystals of HAp orienting along the longitudinal axis of the fibrils.

Another approach was presented in work of Bhowmik and coworkers [233] who performed molecular dynamics (MD) and steered MD (SMD) simulations in order to understand the loadcarrying behaviour of Col in the proximity of HAp. The Col molecule is helical overall with non-helical ends that are N- or C-telopeptides. The Col molecule with telopeptides interacts with specific surfaces of mineralized HAp (Fig. 21).

Simulations indicate that the load-deformation response of Col is different when it interacts with HAp as compared to its response in the absence of HAp. The interface between Col and HAp affects the overall load-deformation response of Col.

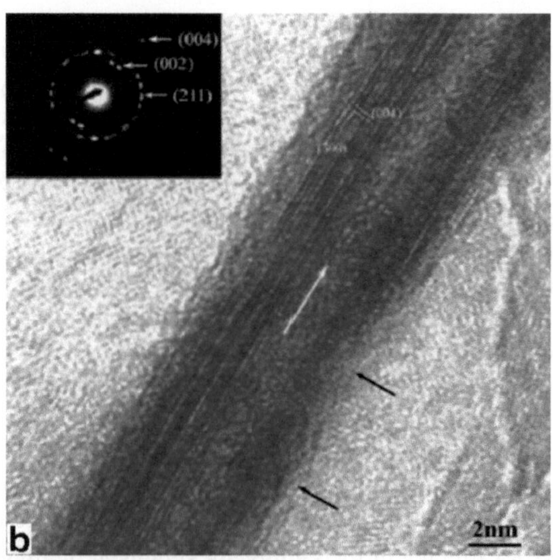

Fig. 20 High magnification of mineralized collagen fibrils. High resolution TEM image of the mineralized collagen fibril; *inset* is the electron diffraction pattern of the selected area. Reprinted from [232] with permission from Elsevier

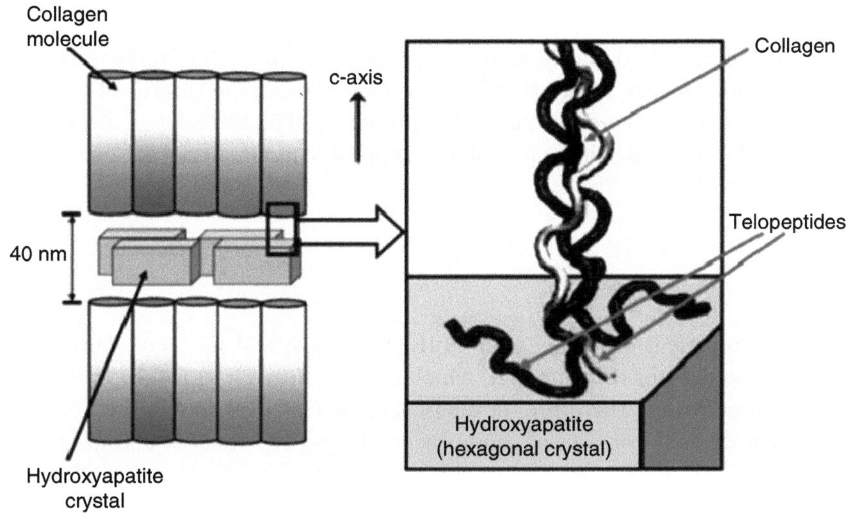

Fig. 21 The Col–HAp interface. Reprinted from [233] with permission from Springer

Further, bone also has a considerable amount of water which significantly influences the load-deformation response of polymer due to specific interactions.

Bernhardt et al. [234] obtained a synthetic material that mimics the composition and structure of the extracellular bone matrix, which mainly consists of ColI fibrils, mineralized with HAp (nano)crystals. This nanocomposite material was produced in a biomimetic process, in which Col fibril assembly and mineralization with HAp occur simultaneously. The authors observed that the membranes from biomimetically mineralized ColI show a substantial influence on the osteogenic differentiation of human bone-marrow-derived stromal cells (hBMSCs). The bone-like composition of the material, combined with its stimulating effect on the osteogenic differentiation of hBMSC, makes it appropriate for human bone regeneration.

The scaffold material nanoHAp/recombinant human-like Col/poly(lactic acid) (nanoHAp/RHLC/PLA), was developed by Wang and coworkers using biomimetic synthesis [235]. The percentages of organic phase and inorganic phase of nanoHAp/RHLC were similar to that of natural bone, and the 3D porous scaffold materials mimic the microstructure of cancellous bone. In the implantation experiment, the segmental defect was healed 24 weeks after surgery, and the implanted composite was completely substituted by new bone tissue. The results of the implantation experiment were very comparable with that of the scaffold based on mineralized animal-sourced Col. It is concluded that the scaffold based on mineralized recombinant Col maintains the advantages of mineralized animal-sourced Col, but avoids potential virus dangers.

Teng et al. [236] prepared a three-layered membrane consisting of Col/HAp and CS in a sandwich structure using a layer-by-layer filtration process. SEM results showed that both the surface and cross-section of the three-layered membranes exhibited a porous structure with a pore size of tens to hundreds of micrometers.

The CS layer acted as a support substrate so as to endow the membranes with a high tensile strength and elastic modulus. At the same time, the Col/HAp composites used as the surface layers provided the membranes with improved flexibility and bioactivity as compared to pure CS membranes. The three-layered membrane prepared in this study would be potentially applicable as a material for guided bone regeneration (GBR).

Gelatin

The biggest practical problems with ColI are its cost and the poor definition of commercial sources of this material, which make it difficult to follow up on well-controlled processing. Therefore, in many studies, ColI was replaced by a gelatin (Gel) precursor. Gel consists of amino acid glycine (33 mol%), amino acid hydroxyproline (10 mol%) and hydroxylysine (0.5 mol%), and shows well-defined physical and chemical properties, even for commercially available lower grade materials [237–239]. Gel/HAp composites have been developed as good candidates for hard tissue repair materials because of their similar composition to the hard tissues, good biocompatibility and high osteoconductive activity [240]. Very recently, they have also been used as a drug delivery system for the treatment of bone infections and defects [241].

Chang and coworkers [237] synthesized Gel/HAp nanocomposites using a biomimetic process. The amount of Gel in a constant volume batch greatly influenced the nucleation and the development of HAp nanocrystals. The coprecipitated Gel/HAp nanocomposites showed chemical bond formation between Gel macromolecules and HAp nanocrystals and had a self-organized structure along Gel fibrils.

A similar method was used by Kim and coworkers [242] to prepare Gel/HAp nanocomposites, which were fabricated as scaffolds by the co-precipitation of HAp within a Gel sol and further freeze-drying. The nanocomposites had a well-developed pore structure, and less-crystallized and finer HAp crystals, which is in marked contrast to the conventionally made composites, Fig. 22.

The human osteoblastic cells attached to the nanocomposites to a significantly higher degree than to the conventional composites. The cells grew and proliferated favourably on the nanocomposites, suggesting that the Gel/HAp nanocomposites obtained biomimetically have high potential for use as hard tissue scaffolds.

The effect of Gel on the nucleation and growth of cHAp crystals and its subsequent influence on the microstructure of the sintered body was examined by Shu et al. [243]. Investigators synthesized cHAp in the presence of Gel molecules. Their experimental results indicated that the concentration of the Gel solution is a predominant factor in inhibition of nucleation and growth. This inhibition is believed to relate strongly to the covalent interaction between Ca^{2+} ions in the solution and the $R\text{–}(COO)^-$ ions of Gel molecules. It also influences the Ca:P ratio, and subsequently affects the microstructure of the sintered body.

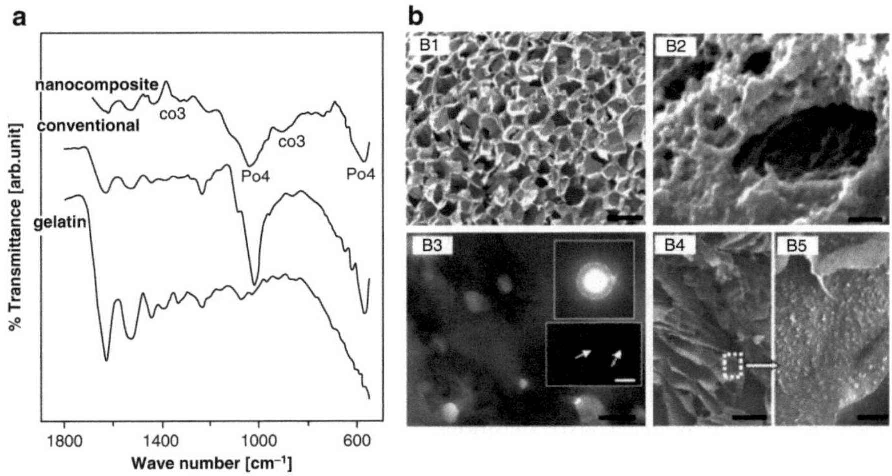

Fig. 22 Characteristics of the gelatin/30% HAp nanocomposites. (**a**) FTIR analysis. Data on gelatin/30% HAp conventional composite and pure gelatin are also presented. (**b1, b2**) SEM and (**b3**) TEM images of the gelatin/30 wt% HAp nanocomposite. Insets in (**b3**) are enlargements showing elongated HA nanocrystals (*lower, arrowed*) and the selected area diffraction pattern of the crystal (*upper*). (**b4, b5**) Images of conventional composite. Scale bars: 500 μm (**b1**), 300 nm (**b2**), 100 nm (**b3**), 20 nm (**b3**, inset), 500 μm (**b4**) and 30 μm (**b5**). Reprinted from [242] with permission from Elsevier

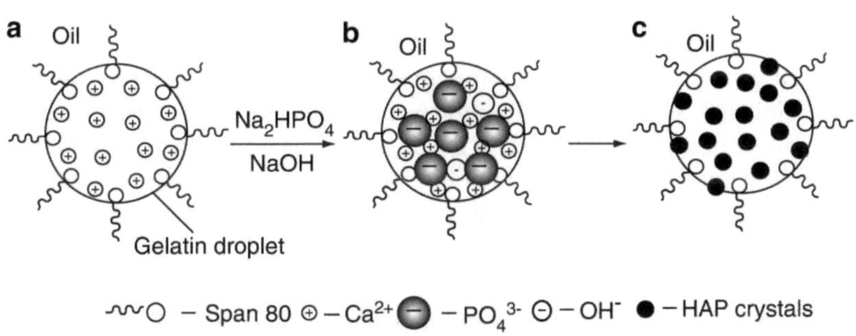

Fig. 23 The growth process of HAp crystals in gelatine droplets. Reprinted from [244] with permission from Elsevier

Interestingly, a one-pot strategy was presented in work of Teng and coworkers [244], who synthesized Gel/HAp composite microspheres in a water-in-oil (W/O) emulsion, as depicted in Fig. 23.

Using Gel droplets as microreactors and colloid protective medium, needle-like nanoHAp crystals (5 nm × 60–100 nm) in the form of clusters were homogeneously and orderly precipitated within the Gel matrix. The results of SEM studies revealed that the as-prepared microspheres with an average diameter of 7.5 μm displayed a narrow particle size distribution, a high dispersity and a naturally porous structure.

Rohanizadeh and coworkers [245] studied three types of scaffolds: Gel sponge (Gelfoam), Gel sponge/HAp, and Gel sponge/PLGA. PLGA coating of Gel sponge increased its biodegradation resistance, but also decreased osteoblast proliferation. Although osteoblast proliferation was slightly enhanced by HAp deposition in Gel sponge, this increase was not significant compared to cell proliferation in the unmodified sponges. It was speculated that changes in sponge structure (smaller pores size), caused by the drying stage after mineral deposition, may compromise the beneficial effects of mineral calcium (HAp) on osteoblast proliferation and adhesion.

Recently, Hillig and coworkers [246] reinforced the Gel matrix with fine HAp and looked at its suitability as a surgical implant. They synthesized Gel/HAp composites over a range of compositions having an open porous mesoscale structure, and determined the mechanical properties and swelling behavior of such composites upon exposure to quasi-vivo conditions. Their results show a strong dependence of the properties on composition and on whether the materials had been soaked in a body-simulating normal saline solution. The least swelling and the greatest stiffness occurred at the greatest (feasible) Gel/HAp ratios. The decrease in the compressive stiffness for most compositions after imbibing moisture could be related to the significant void content left behind as the gelatin gel dries. Such empty space would provide room into which the hydrated Gel matrix could expand without imposing a large counter stress on the HAp constituent [246].

In another work, an in situ biomimetic synthesis of Gel/HAp nanocomposite powders with varied proportions was described [247]. The use of HAp crystals and Gel, the soluble form of bone protein, makes the nanocomposites comparable to natural bone in constituents. The authors found that the application of biomimetic principles improves crystal morphology and the interaction of HAp crystals with the Gel molecules, as seen through in vitro characterizations. The best results were obtained for 80:20 proportions of HAp to Gel, which proved to be closest to the characteristics of natural bone. The in vivo implantation studies in the femoral condyle of the animals, as assessed by serial post-operative follow-up radiography and histological evaluation, revealed a good biocompatibility and bone-regeneration ability of the material.

Catledge et al. [248] prepared Gel/nanoHAp composite scaffolds with up to 40 wt% of HAp using an electrospinning technique. Dispersion of nanoHAp particles in the fibres was good up to about 30 wt%, but agglomeration of HAp particles became pronounced above this concentration. Nanoindentation of uniaxially pressed scaffolds revealed very high Young's modulus values ranging from 5.9 ± 0.9 GPa (pure Gel) to 10.2 ± 0.8 GPa (Gel with 40 wt% nanoHA). The high elastic modulus values can be explained by the scale of measurement – at smaller length scales, measurement approaches that of the molecular level, and is therefore independent of factors such as porosity.

Silk

Silk fibroin (SF) polymers consist of repetitive protein sequences and provide structural roles in cocoon formation, nest building, traps, web formation and egg protection. Silks are generally composted of β-sheet structures due to the dominance of hydrophobic domains consisting of short sidechain amino acids in the primary sequence. Silk is biocompatible, degradable and shows superior mechanical properties. Silk materials are amenable to aqueous or organic solvent processing and can be chemically modified to suit a wide range of biomedical applications [249–251].

Nemoto and coworkers prepared HAp nanocrystals at room temperature by a coprecipitation method from $Ca(OH)_2$ and H_3PO_4, in the presence of chemically disintegrated SF [252]. Adsorbed amounts of cations on SF and the crystallinity of HAp in the composite were increased by the chemical disintegration of SF higher-order structure. Preferential alignment of the *c*-axis of HAp crystallites along the longitudinal direction of an ca. 150-nm SF fibril was observed. These changes due to disintegration of SF were discussed in terms of the chemical interactions between SF and HAp, as studied by FTIR. Disintegration of SF induces a larger number of nucleation sites for chemical interactions between HAp and the C=O of disintegrated SF fibrils [252].

Multilayered SF/HAp films were prepared by Kino et al. [253] using SF, HAp as the main component and methanol-untreated SF as the binder during alternating lamination. The cross-sectional morphology showed macroscopic uniformity of HAp in SF. The β-sheet content of SF increased at a lamination temperature of 130°C, contributing to the bonding strength between the individual films. The uniform structure, good osteocompatibility and high mechanical strength of the multilayered SF/HAp films suggest their potential for bioabsorbable membranes.

In the work of Korematsu and coworkers [254], composites consisting of SF and nanoHAp particles were obtained according to the synthetic route presented in Scheme 2.

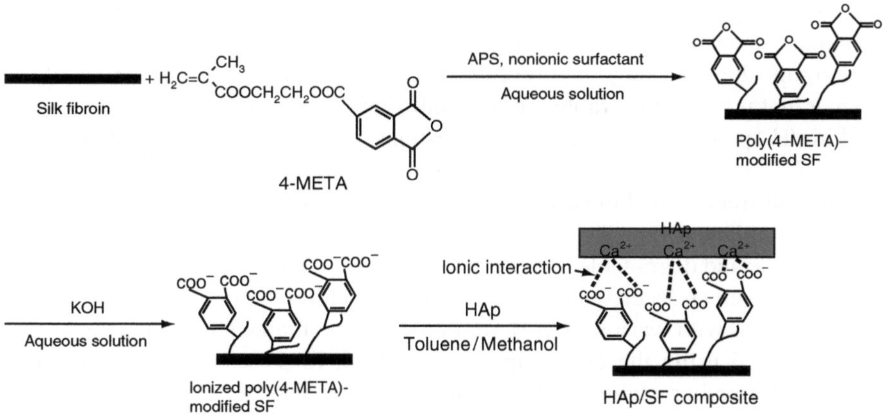

Scheme 2 Synthesis of the SF/HAp composite. Reprinted from [254] with permission from Springer

HAp nanoparticles were adsorbed equally and dispersively on the treated SF fibre surface by ionic interactions. This synthetic system requires no heat to connect HAp to SF and is useful when applying to non-heat-resistant biomedical materials. The cell-adhesion test shows that the SF/HAp composites improves bioactivity compared to the original SF, and that they do not cause inflammation in living bodies.

7.2.2 Polysaccharides

Polysaccharides contain monosaccharide units joined together by glycosidic linkages. Polysaccharides display unique biological functions ranging from cell signalling to immune recognition. This, combined with new synthetic routes currently available for synthesis or modification of polysaccharides, biodegradability and the ability to form specific interactions make them one of the most important and extensively investigated groups of natural biomaterials.

Chitin and Chitosan

Chitin and CS are naturally occurring copolymers of *N*-acetyl-glucosamine and *N*-glucosamine; the comonomer units are randomly or block distributed, depending on the processing method used to derive the biopolymer. When the number of *N*-acetyl-glucosamine units is higher than 50%, the biopolymer is termed chitin. Conversely, when the number of *N*-glucosamine units is lower, the term chitosan is used. CS has been the better researched biopolymer because of its ready solubility in dilute acids, rendering CS more accessible for utilization and chemical reactions. Chitin and CS are important constituents of the exoskeleton of crustacea, molluscs and insects. The main sources exploited are two marine crustaceans, shrimp and crab [255–259].

Wan et al. have investigated the interactions of chitin with calcium species [257]. In one strategy, HAp was dispersed in chitin to produce intimately blended material. Preliminary mechanical tests revealed a reduction in strength for the more highly filled composites, but they also revealed retention of the plastic properties of the polymer that may be favourable for bone substitute applications [260].

Chitin/HAp composites were also investigated by Ge et al. [261]. HAp in 25, 50, and 75% (w/w) fractions was incorporated into chitin solutions and processed into air- and freeze-dried materials. These materials were then exposed to cell cultures and implanted into the intramusculature of a rat model; they proved to be non-cytotoxic and degradable in vivo. The presence of the HAp filler enhanced calcification as well as accelerated degradation of the chitin matrix.

Composites with various CS/HAp ratios were obtained by Yamaguchi and coworkers [45] using the co-precipitation method. In these composites, calcium phosphate formed crystalline HAp when acetic acid and lactic acid were used in the preparation solvents for CS. The calcium phosphate was found to be amorphous when organic acids having more than two carboxyl groups were applied. Results of their investigations indicated that CS/HAp composites consist of homogeneous aggregations 230 nm in length, in which many HAp nanocrystals align along the CS molecules. The authors suggested that the HAp nanocrystals grew on nucleation sites; that is, the amino groups in CS formed a complex with calcium. The composites were found to be mechanically flexible and could easily be formed into any desired shape. The mechanical strength could be enhanced by heat treatment in a saturated steam, and this effect was ascribed to the formation of hydrogen bonds between CS macromolecules.

Biodegradable CS/Gel/HAp composites were prepared by Zhao et al. [262]. They obtained similar compositions to that of normal human bone as 3D biomimetic scaffolds by phase separation. By changing the solid content and the compositional variables, the authors controlled the porosities and densities of the scaffolds. Histological and immunohistochemical staining and SEM observations indicated that the osteoblasts attached to and proliferated on the scaffolds. The presence of HAp in the CS/Gel composite promoted initial adhesion of human mesenchymal stem cells (hMSC) and supported long-term growth in 3D porous CS/Gel/HAp scaffolds [263].

In the work of Li et al. [264], a series of CS/Gel/nanoHAp composites has been fabricated by nanoHAp depositing on the surface of CS/Gel network films as presented in Fig. 24.

The amount of polymers in the CS/Gel networks greatly influenced nucleation and the development of the nanoHAp crystalline phase. The $-COO^-$, $=C=O$ or $-NH_2$ groups can be active sites for the coordination of calcium ions to form ion complexes, which initiate and control formation of nanoHAp nuclei. One can therefore control the size of nanoHAp through changing the ratio of CS and Gel to imitate the structure of natural bone.

Zhang et al. [265] obtained CS/HAp composites by a co-precipitation method. An excellent miscibility and no phase-separation between CS and nanoHAp was observed due to strong specific interactions between the components, which endows these composites with good mechanical strength. The maximum compressive strength was about 120 MPa for the 30/70 wt% CS/nanoHAp composite. In vitro tests showed that degradation of CS in this composite takes place and that a layer of bone-like apatite forms on the surface of the composite – a sign of its high bioactivity.

The issue of the mechanical resistance of CS-based composites with HAp was addressed by Hu et al. [266]. Investigators reported CS/HAp multilayer nanocomposites with high strength and bending modulus, rendering the materials suitable for possible application in internal fixation of long bone fractures.

Murungan and coworkers [267] demonstrated the possibilities of CS/HAp nanocrystal composite production using a wet chemical method at low temperature. Using this approach, some of the disadvantages of nanoHAp, such as bioresorption and particle migration, can be alleviated. The rate of bioresorption of nanoHAp was improved by the addition of CS. The viscoelastic nature of the composites would prevent the migration of particulate matter into the surrounding tissue upon post-implantation. Besides, the smoothness of the composites would not damage or harm any soft tissue around the implant.

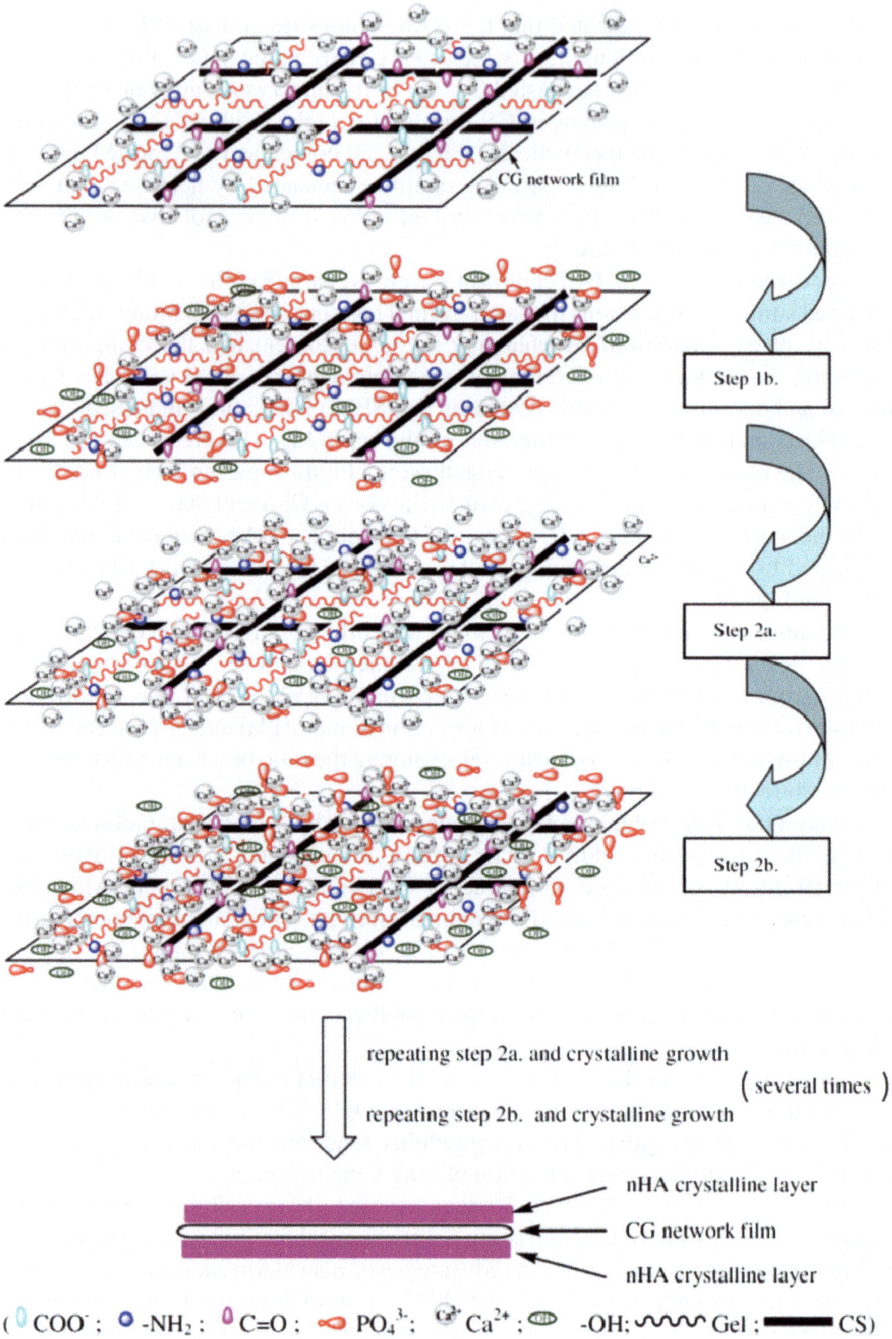

Fig. 24 Mechanism of nanoHAp crystalline formation in situ on the surface of CS/Gel network films. Reprinted from [264] with permission from Elsevier

Injectable bone substitute material consisting of CS, citric acid and glucose solution as the liquid phase, and tricalcium phosphate powder as the solid phase, was developed by Liu and coworkers [268]. Researchers used four groups of cements to investigate the mechanical properties and in vitro biocompatibility of the material. Tricalcium phosphate partially transformed into HAp and dicalcium phosphate in the presence of citric acid.

Kong et al. [269] have investigated CS/nanoHAp scaffolds for bone tissue regeneration. They studied the bioactivity of the composite scaffolds by examining the apatite formed on the scaffolds incubated in SBF, and the activity of pre-osteoblasts cultured on them. After incubation in SBF on both of the scaffolds (CS and CS/nanoHAp) cHAp was formed, furthermore, with increasing nanoHAp content in the composite, the quantity of the apatite formed on the scaffolds increased. The authors suggested that compared with pure CS, the composite with nanoHAp could form apatite more readily during the biomimetic process. It is an important finding because cells presented better proliferation on the apatite-coated scaffolds than on CS scaffolds.

CS/nanoHAp microsphere-based scaffolds were produced by Chesnutt et al. [270] using a co-precipitation method. The authors found that surface area and surface roughness of composite scaffolds were significantly greater than that of scaffolds composed of only CS. Moreover, composite scaffolds swelled significantly less than CS scaffolds and are therefore expected to maintain their shape better in vivo. The compressive modulus of composite scaffolds was higher than the modulus of CS scaffolds, and both were higher than previous CS scaffolds fabricated by other techniques. Osteoblast proliferation was significantly increased on composite scaffolds compared to CS scaffolds.

Recently, Kim et al. [271] obtained a highly flexible HAp/CS composite film using homogeneously dispersed solutions of microscaled HAp powder and CS at various concentrations. These films can contain up to 70% HAp without microcracks, phase separation and brittleness. The results showed that the HAp powders were well-dispersed in the film without clustering or aggregation. These observations were explained by the presence of CS that retains the physico-mechanical properties of the film and captures the HAp powders in the matrix. It is anticipated that these films will be potentially useful for the direct or indirect delivery of HAp materials in localized areas of osteoporosis via a patch on skin or bone.

Hybrid CS/HAp composite materials were also developed by Araujo et al. [272]. The effects of the concentration of a cross-linking agent (genipin) and of lactic acid, and of the presence of HAp powder on the evolution of rheological properties were studied. It was found that the concentration of lactic acid has a marginal influence on the rheological properties, affecting mostly the final microstructure, with pore sizes decreasing with increasing concentration of lactic acid. The microstructure of CS scaffolds generally consists of large and interconnected pores, the size of which also tends to decrease with increasing amounts of genipin. Moreover, the CS and CS/HAp scaffolds exhibited an excellent bioactivity in vitro in the presence of SBF, showing the formation of a fair apatitic layer, even in the early stages of immersion.

Starch

The starch molecule has two important functional groups, the –OH groups that are susceptible to substitution reactions, and the C–O–C groups where chain breakage starts. By reaction of its –OH groups, chemical modification of starch can be performed, leading to materials with various properties [66].

Biodegradable starch-based polymers have recently been proposed as having great potential for several applications in the biomedical field, such as bone replacement implants, bone cements, drug delivery systems, and tissue engineering scaffolds [273]. The development of new processing techniques and the reinforcement with various (nano)fillers has resulted in materials with mechanical properties matching those of bone [274].

Mendes and coworkers [275] described an extensive biocompatibility evaluation (in vitro and in vivo) of biodegradable starch-based materials aimed at orthopaedic applications as temporary bone replacement/fixation implants. For that purpose, they studied a polymer starch/ethylene vinyl alcohol blend (SEVA) and a composite of SEVA reinforced with HAp particles. As a result of their investigation it was found that SEVA and SEVA/HAp materials did not show relevant toxicity in both short- and long-term in vitro testing. Furthermore, when implanted in muscle or bone tissue, these materials induced a satisfactory tissue response.

The biocompatibility of two different blends of corn-starch, SEVA and starch/cellulose acetate (SCA), and their respective composites with HAp, were studied by Marques et al. [274]. Researchers found that both types of starch-based polymers exhibit a cytocompatibility that might allow for their use as biomaterials Furthermore SEVA blends were found to be less cytotoxic for the tested cell line, although cells adhere better to SCA surface. Considering the overall behaviour of SEVA, SCA and their composites with HAp, it can be expected that their cytocompatibility will allow for their use in the future in applications such as bone replacement/fixation and/or tissue engineering scaffolding.

Cellulose

Cellulose differs in some respects from other polysaccharides produced by plants, the molecular chain being very long and consisting of one repeating unit (cellobiose). The most popular method of cellulose production is isolation of cellulose from plants, but the principal pathways of cellulose preparation include biosynthesis by different types of microorganisms, enzymatic in vitro synthesis, and chemosynthesis from glucose derivatives. The properties of the bacterial cellulose are quite different from those of plant celluloses especially concerning the ultrafine network architecture, high hydrophilicity, and mouldability during formation [276, 277]. In all forms, cellulose is a very highly crystalline, high molecular weight polymer, which is infusible and insoluble in all but the most aggressive, hydrogen-bond-breaking solvents such as *N*-methylmorpholine-*N*-oxide. Because of its infusibility and insolubility, cellulose is usually converted into derivatives to make it more processable [66].

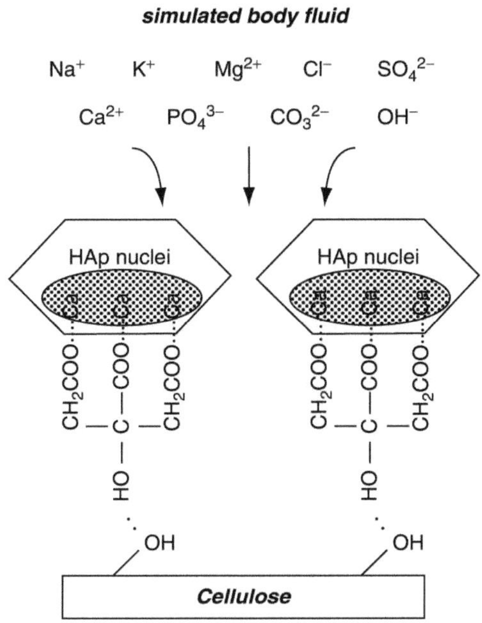

Fig. 25 Formation of HAp embryo with the aid of citric acid in SBF. Reprinted from [278] with permission from Springer

Rhee and coworkers [278] studied HAp formation on cellulose with the aid of citric acid. Cellulose was soaked in SBF solutions (1.5 SBF) with ion concentrations 1.5 times that of SBF (1.0 SBF) with and without citric acid carbonate containing HAp crystals were found to form only in the 1.5 SBF solution that contained citric acid. The results were explained in terms of hydrogen bonding of citric acid to cellulose and its ability to chelate calcium ions (Fig. 25).

Bacterial cellulose (BC)/HAp nanocomposites were examined by Wan et al. [279]. The most striking features of BC are its high mechanical strength and modulus, as well as its biodegradability. Compared with other natural biodegradable polymers, BC presents much better mechanical properties, which are required in most cases when used as scaffold in tissue engineering. Compared with animal-derived polymers, BC is free of any occurrence of cross-infection that can be associated with collagen [276]. The authors found that there are different interactions between unphosphorylated and phosphorylated BC fibres and HAp, as shown schematically in Fig. 26.

It was found that phosphorylated BC could act as potential substrate for apatite nucleation. The observed 3D porous network structure and interconnected pores, which can be adjusted over a wide range, make the BC/HAp composites promising materials in tissue engineering. In addition, it is expected that the advantageous mechanical properties of BC will allow the design of a wide range of BC/HAp composites with mechanical properties ranging from those analogous to soft tissues to those similar to hard tissues by controlling the ratio of HAp to BC and their 3D structure.

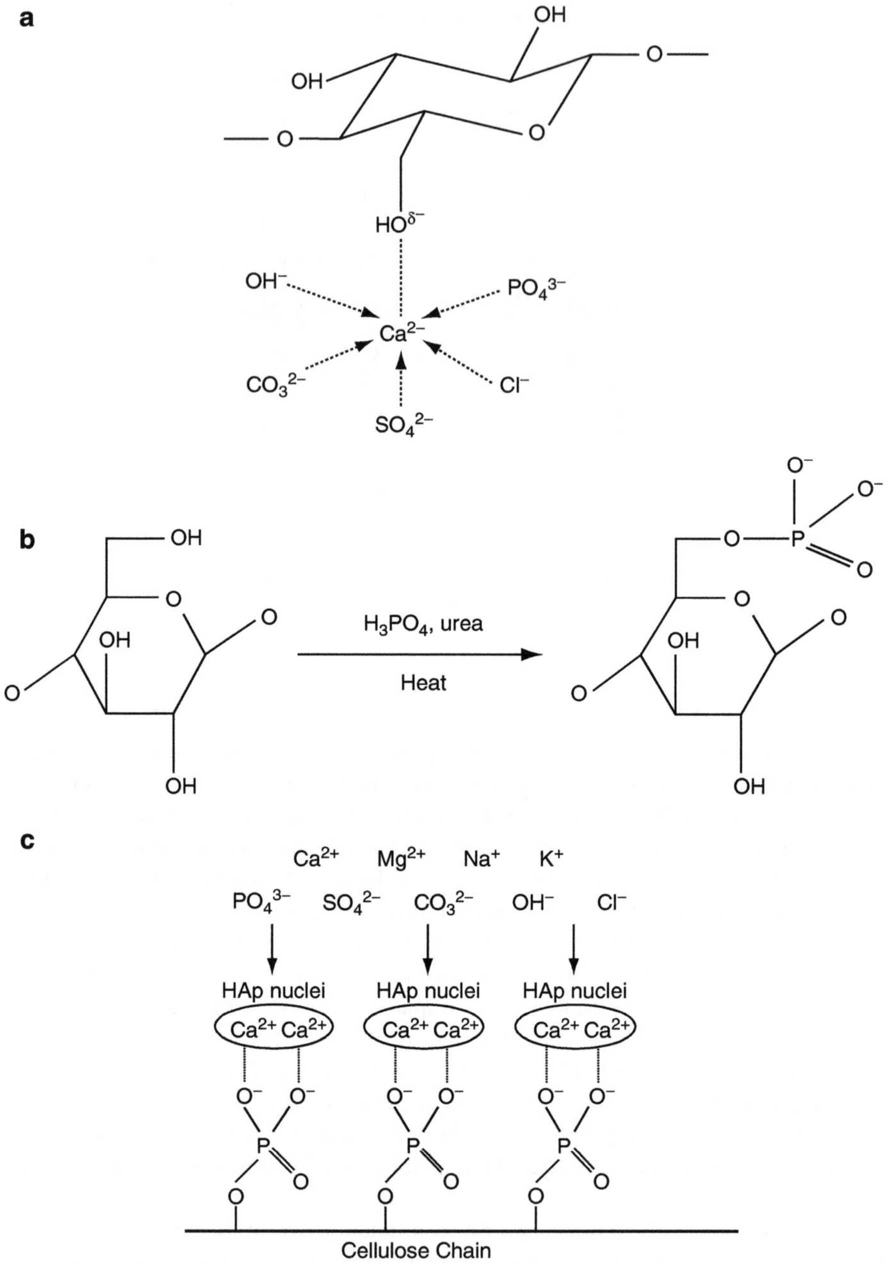

Fig. 26 Various interaction processes between BC and HAp. (**a**) HAp growth on unphosphorylated BC; (**b**) phosphorylation of BC; (**c**) HAp growth on phosphorylated BC. Reprinted from [279] with permission from Elsevier

Jiang and coworkers [280] developed a new type of bone-replacing material composed of different weight ratios of nanoHAp and a CS-carboxymethylcellulose network. NanoHAp was uniformly dispersed in the composite in the form of nanometer-grade short crystals, which ensured that the composite had high compressive strength; for the composite with 40 wt% the compressive strength reached nearly 120 MPa, which can meet the requirement of initial mechanical properties for bone repair material. Moreover, its weight loss was up to 56.44% after soaking in SBF for 8 weeks, which indicates a degradable composite. Next, apatite particles aggregated to form a bioactive apatite layer deposited on the surface.

Alginic Acid

Alginate (Alg) is a binary linear heteropolymer containing 1,4-linked α-L-guluronic acid and β-D-mannuronic acid. Alginates are extracted from the algae using a base solution and then reacted with an acid to form alginic acid. They are high molecular weight polymers having molecular weights of up to 500 kDa. Alginic acid forms water-soluble salts with monovalent cations, low molecular weight amines and quaternary ammonium compounds. It becomes waterinsoluble in the presence of polyvalent cations such as Ca^{2+}, Be^{2+}, Cu^{2+}, Al^{3+} and Fe^{3+} [60, 66]. The high functionality of alginic acid makes it a favourable biopolymeric material for use in medical applications [281, 282].

Zhang et al. [282] synthesized porous Col/Alg/nanoHAp composites containing Ca-cross-linked Alg. These composites showed a significant improvement in mechanical properties over Col/nanoHAp materials. Mechanical test results show that the compressive modulus and yield strength are in direct proportion to the percentage of Ca-crosslinked Alg in the composite. Primary in vitro biocompatibility experiments, including analysis of fibroblast and osteoblasts formation, indicated the high biocompatibility of these composites.

In another work, Stodolak et al. [284] investigated three composite materials consisting of poly(D,L-lactide) matrix (PDLLA) and Alg fibrous phase with nanoHAp: PDLLA/NaAlg/nanoHAp, PDLLA/$Ca(Alg)_2$/nanoHAp and PDLLA/NaAlg/$Ca(Alg)_2$/nanoHAp. The authors found that the introduction of Alg fibres modified with nanoHAp into the bioresorbable polymer matrix influences the degradation process of the composites. Alg fibres may accelerate NaAlg/nanoHAp degradation or retard $Ca(Alg)_2$/nanoHAp degradation. The composites can be used for fabrication of porous materials with a controlled degradation time that can be used in GTR/GBR techniques.

Recently, Wang and coworkers [285] prepared a series of HAp/Alg nanocomposites with Alg amounts varying from 10 to 40 wt% using in situ hybridization. As expected, the crystallinity of HAp decreased with the increase of Alg content. FTIR spectroscopy results indicated that the chemical interaction occurred between the mineral phase (Ca^{2+}) and the polymer matrix (carboxyl groups). As compared to pure HAp without Alg, the composites showed more homogeneous (micro)structures, where HAp nanocrystals were well-embedded in the Alg matrix. The composite containing 30 wt% Alg exhibited a highly ordered 3D network, similar to natural bone structure.

Hyaluronic Acid

Hyaluronic acid (HYA) is a member of the glycosaminoglycan family, which are linear polysaccharides consisting of alternating units of *N*-acetyl-D-glucosamine and glucuronic acid, and are found in virtually every tissue in vertebrates. HYA is water-soluble and forms highly viscous solutions with unique viscoelastic properties. It can form 3D structures in solution with extensive intramolecular hydrogen bonding [60, 286, 287].

Solchaga and coworkers [288] investigated the possibility of using HYA-based polymers as cell carriers for tissue-engineered repair of bone and cartilage. Their results indicated that HYA-based delivery vehicles are superior to porous calcium phosphate ceramics with respect to the amount of cells loaded per unit volume of implant. Weak mechanical strength to give sufficient structural support and protection for the seeded osteoblasts, and the risks of pathogen transmission and immunorejection associated with natural materials from animal and cadaver sources, are properties that still need to be improved for composite scaffolds applications [48].

Composites made from Col and HAp conjugated with HYA were studied by Bakos and coworkers [289]. Results of investigations showed more dense composite arrangements due to the formation of Col–HYA conjugates, and that particles of inorganic component were closely anchored in the structure. The mechanical properties and swelling behaviours of these composites were compared with those of Col/HAp composites of similar material without HYA. Besides, Kim and Valentini [290] have demonstrated that benzyl-derivatized HYA formed into porous scaffolds can support in vivo bone ingrowth.

8 3D Polymer/HAp Scaffold Design and Fabrication

Design and fabrication of tissue-engineering scaffolds are currently an important research theme in biomedicine The structure of the scaffold should act as a template for 3D tissue growth; it consists of a highly interconnected porous network with interconnections greater than 50 μm in diameter and pore diameters larger than 100 μm [290, 292].

Sufficient porosity is required to accommodate osteoblasts or osteoprogenitor cells, to support cell proliferation and differentiation, and to enhance bone tissue formation. High interconnectivities between pores are also desirable for uniform cell seeding and distribution, the diffusion of nutrients to and metabolites out from the cell/scaffold constructs. The scaffold should have adequate mechanical stability to provide a suitable environment for new bone tissue formation, as well as appropriate surface structure for adhesion and function of bone cells [48]. These requirements led to development of polymeric scaffolds reinforced with HAp.

Porous 3D polymeric scaffolds with HAp for bone regeneration are fabricated by solvent casting and particulate leaching, melt moulding, emulsion freeze drying, gas foaming, electrospinning, thermally induced phase separation or solid freeform fabrication (SFF) (Table 3).

8.1 Solvent Casting and Particulate Leaching

Solvent casting and particulate leaching (SC/PL) is a simple and commonly used method for fabricating scaffolds for tissue engineering. With an appropriate thermal treatment, porous constructs of synthetic biodegradable polymers can be prepared with specific porosity, surface:volume ratio, pore size and crystallinity for different applications. This method involves mixing water-soluble salt (e.g. sodium chloride, sodium citrate) particles into a polymer solution. The mixture is then cast into the mould of the desired shape. After the solvent is removed by evaporation or lyophilization, the salt particles are leached out to obtain a porous structure with the pore shape limited to the cubic crystal shape of the salt. Removal of soluble particles from the interior of a polymer matrix is difficult and limits the thickness of the films prepared to ca. 2 mm [293].

8.2 Melt Moulding

Melt moulding, an alternative method of constructing 3D polymer/HAp scaffolds, has many advantages that are offered by modern processing technologies, such as reproducibility, shapeability, homogeneous distribution of filler and low cost.

In melt moulding, special attention has been given to the optimization of the processing techniques (and respective processing parameters), including compounding, compression and injection moulding and extrusion, and to the selection and characteristics of the reinforcement (composition, shape, granulometric distribution) [78]. The technique is particulary effective if twin-screw extrusion is used in processing of polymer composites since large shear forces cause unifiorm particle distribution [82, 294]. Numerous examples of polymer/HAp composites prepared by melt moulding techniques are given in Tables 1–3.

8.3 Emulsion Freeze-Drying

Another method for fabricating polymer/HAp scaffolds with variable porosity and pore size utilizes an emulsion freeze-drying process. In this technique, control of

Table 3 Methods of fabrication and characteristics of polymer/HAp scaffolds

Method	Polymer/HAp scaffolds	Pore size	Morphology	Comments	Ref.
Solvent casting and particulate leaching	PLGA/HAp	Gelatine: 23–52 μm Salt: 67–167 μm	Open-cell	Porogens: either gelatin microspheres or NaCl salt particles	[154]
Melt moulding	PE/HAp	–	–	Compounding in the extruder, granulating and injection moulding	[71–74, 76–84]
	POM/HAp	–	–	Compounding in the extruder, granulating and injection moulding	[99, 100]
	PSU/HAp	–	–	Compounding in the extruder, granulating and injection moulding	[102–105]
	PEEK/HAp	–	–	Compounding in the extruder, granulating and injection moulding	[11, 108, 109]
	PHB/HAp	–	–	Compounding, milling, and compression moulding	[170, 173]
	PCL/HAp	–	–	Conventional blending in an extruder	[166]

(continued)

Table 3 (continued)

Method	Polymer/HAp scaffolds	Pore size	Morphology	Comments	Ref.
	SEVA/HAp	–	–	Processed by a non-conventional injection moulding technology, SCORIM to induce a high anisotropic level into the mouldings	[275]
Emulsion freeze-drying	Gel/HAp	400–500 μm	Closed pores, well-developed pore structure	Fabricated by the co-precipitation of HA within a gelatin sol and further freeze-drying; porosity 89–91%	[242]
	Chitin/HAp		Closed pores	HAp fraction was incorporated into chitin solutions and processed into air- and freeze-dried materials	[261]
Gas foaming process	PLA/HAp PLA/β-TCP	0.40–0.07 mm	Open pores	Supercritical CO_2, ceramic fillers and polymer were first mixed before the foaming process by melt extrusion anisotropy,	[148]

(continued)

Table 3 (continued)

Method	Polymer/HAp scaffolds	Pore size	Morphology	Comments	Ref.
				both in morphology and in mechanical properties	
Electrostatic spinning	PLA/PLA-*g*-HAp		Composite nanofibres	Diameter of PLA-*g*-HAp/PLA fibres from 1.10 μm for PLA to 600 nm for PLA-*g*-HAP (30 wt%), porosity 88–89%	[149]
Phase separation	CS/Gel/HAp	300–500 μm	Open pores	Solid–liquid phase separation and subsequent sublimation of solvent; porosity 85–95%	[262]
Solid freedom fabrication	Col/HAp	44–135 μm		Phase change jet printer; porosity 87–95%; SSF has the ability to create internal channels of complex geometries within a collagen-HAp scaffold	[302]

the crystal growth of ice is crucial for obtaining a suitable pore diameter and shape, because the pore structure is a replica of the entangled dendrites of ice crystals. The pore diameter can be controlled by freezing temperatures, and pore interconnectivity can be regulated by the concentration of solute [295, 296]. Recently, Yunoki et al. [297] used this method to obtain a 3D porous structure of Col/HAp composite. The freezing apparatus for unidirectional growth of ice crystals and Col/HAp composites structure is shown schematically in Fig. 27.

By using this technique it was possible to obtain the Col/HAp composites with random and spherical pores, as well as to introduce cross-linkages among the Col molecules, enhancing subsequent formation of tubular micropores.

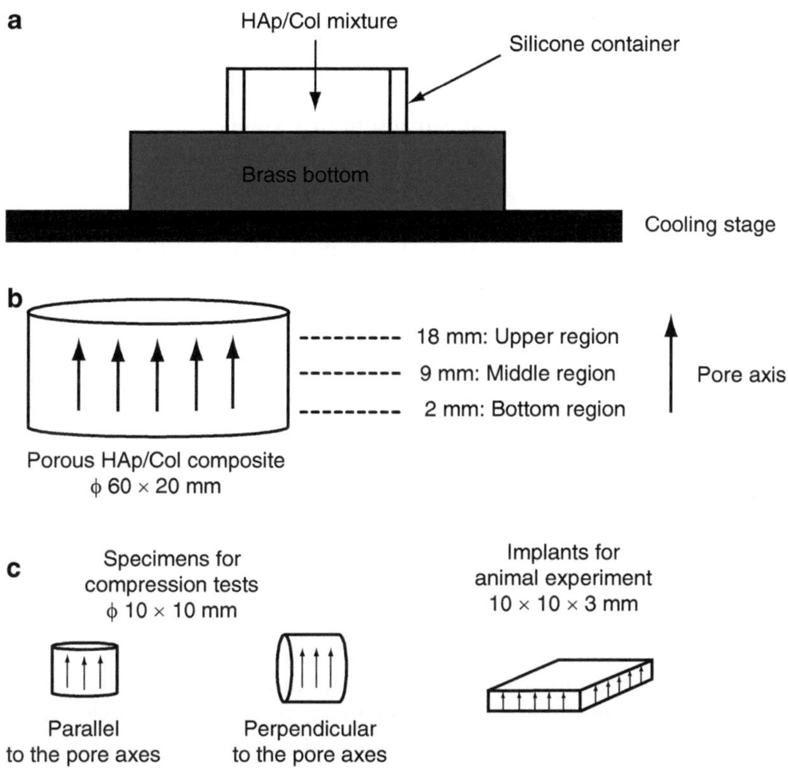

Fig. 27 Schematic drawing of (**a**) the freezing apparatus for unidirectional growth of ice crystals; (**b**) the porous Col/HAp composite obtained and the sites for SEM observation; and (**c**) specimens for the mechanical tests and animal experiments. The container was composed of two parts: a brass bottom with high thermal conductivity, and silicone sides. Only the brass bottom was cooled to generate ice growth along the vertical direction. Reprinted from [297] with permission from Wiley InterScience

8.4 Gas-Foaming Processes

Gas-foaming has been developed as a fabrication method for porous polymer scaffolds and avoids the use of organic solvents. This is desirable because their residues can have toxic effects in vitro and elicit an inflammatory response in vivo. In the gas-foaming method the moulded samples are exposed to high pressure CO_2 to saturate the material. The subsequent reduction in pressure to ambient levels causes the nucleation and formation of pores in the polymer matrix from the CO_2 gas [293].

Several research groups have reported batch foaming of various polymers including biodegradable materials Foaming is mainly applicable to tissue engineering, tissue regeneration and protein encapsulation. Good control over porosity was possible in the reported studies by manipulation of temperature and pressure. The cell number density increased and the cell size decreased with increasing pressure and decreasing temperature. A high degree of supersaturation of dissolved CO_2 at high pressure and low temperature are responsible for such results [298].

The disadvantage of this method is that it mostly yields a nonporous surface and closed-pore structure, with only 10–30% of interconnected pores. The porosity and interpore connectivity can be significantly improved by combining particulate leaching with the gas-foaming process, although completely eliminating closed pores remains challenging [48]. As an example, Mathieu et al. [148] recently obtained open-cell composite foams made of bioresorbable PLA and ceramic fillers, HAp or β-TCP by using supercritical CO_2 foaming (see Sect 7.1.1).

8.5 Electrostatic Spinning

Electrostatic spinning (or electrospinning) is an interesting method for producing nonwoven fibres with diameters in the range of submicrometers down to nanometers. In this process, a continuous filament is drawn from a polymer solution or melt through a spinneret by high electrostatic forces and later deposited on a grounded conductive collector [299].

Electrospinning is increasingly being used to produce fibres for tissue culture scaffolds, which exhibit important advantages when compared with foams. Firstly, the interconnectivity of voids available for tissue ingrowth is perfect, whereas in the case of foams some cavities can be dispersed in the matrix and hence closed; also fenestration between adjacent cavities can be too small to allow for cell permeation. Secondly, ultrathin fibres produced by electrospinning offer an unsurpassed surface:volume ratio among applied tissue scaffolds [300, 301].

8.6 Phase Separation

In this method, the polymer is dissolved in a solvent at a low temperature. Homogeneous solution is then cooled in a controlled fashion until liquid–liquid phase

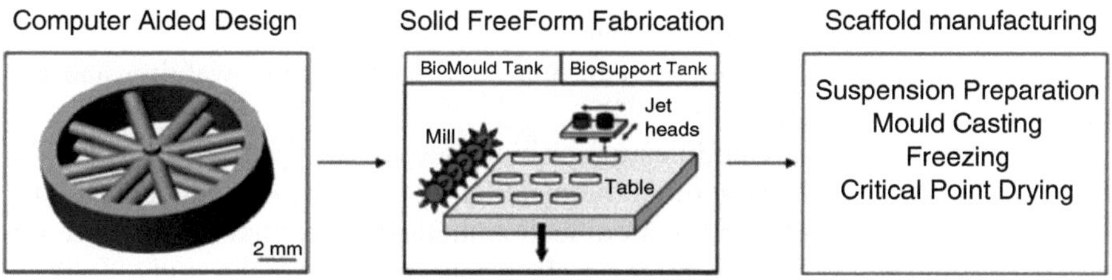

Fig. 28 The indirect process of scaffold engineering with solid freeform fabrication using a dual phase change jet printer (T66, SolidScape, Merrimack, NH). Reprinted from [302] with permission from Springer

separation is induced. The resulting bicontinuous polymer and solvent phases are then quenched to create a two-phase solid. The solidified solvent is then removed by sublimation, leaving a porous polymer scaffold [293].

Phase separation has been utilized to produce scaffolds mostly for drug delivery purposes, but they often have good mechanical properties and may find use as implants. For example, a PLLA scaffold fabricated using a solid–liquid phase separation technique has a Young's modulus approximately 20 times higher than that of scaffold fabricated using the salt-leaching technique from the same polymer and with the same porosity. However, phase separation usually generates scaffolds with a pore size of 10–100 μm, which may not be ideal for osteoblastic cell seeding and bone tissue growth [48].

8.7 *Solid Freeform Fabrication*

Solid freeform fabrication (SFF) is a relatively recent technique used in bone tissue engineering. 3D printing technologies have the potential to create polymer scaffolds directly or indirectly (via mould casting) using a computer-aided design, with predefined external and internal morphologies. Figure 28 shows the indirect process for creating a composite scaffold of pre-designed shape using SFF [302].

Combined with medical scan data from computerized tomography or magnetic resonance imaging, SFF has been used to generate scaffolds that match osseous defects very precisely. It has been applied to produce polymer/HAp scaffolds with porosity, pore size and shape that optimize the mechanical properties [303] or with a design that facilitates the seeding process and enhances cellular adhesion and proliferation under dynamic conditions [304].

9 Surface Modification of Polymeric Implants

The response of the host organism at the macroscopic, cellular and protein levels to biomaterials is, in most cases, closely associated with the material's surface properties. In tissue engineering, regenerative medicine and many other biomedical fields, surface engineering of bioinert synthetic polymers is often required to introduce bioactive species that can promote cell adhesion, proliferation, viability and enhanced ECM secretion functions [305].

The strategy of surface modification of biomaterials has been adopted over the years in order to alter the area of the biomaterial that first comes into contact with a biological environment. Surface modification methodologies have been used in a variety of applications, with many researchers concentrating on the study of different surface stimuli in order to optimize the short-term and long-term performance of biomaterials [306–308]. Crucial factors for surface biocompatibility are wettability (hydrophilicity/hydrophobicity), chemistry, charge and roughness [309–311]. These can be controlled by applying physical methods (e.g., adsorption), chemical methods (e.g. grafting) and other methods such as plasma or corona treatment.

9.1 Physical Methods

The main methods of physical surface modification of polymeric materials are adsorption via hydrogen bonding, hydrophobic bonding, van der Waals force, electrostatic interactions and ligand–receptor pairing. Physical adsorption requires the incubation of the adsorbent with a stationary surface. For gradient formation, an adsorbent might be hydrogel, block copolymer, fibronectin, etc., dissolved in various solvents. Polymer marix bearing functional groups can be used as a template for the second reaction step to introduce other biomolecules or even biomaterial grafts. As a disadvantage, the technique can only be applied to hydrophilic inorganic substrates such as silicon, quartz, glass or HAp [309]. Physical adsorption and self-assembly, which do not require chemical processing, are unfortunately dependent on unpredictable, nonspecific interactions between the protein and the surface [312, 313].

9.2 Chemical Methods

The use of a chemical reaction on a polymer surface is an important method for preparing a gradient surface. The key advantage of this technique is that the surface can be modified to have very distinctive properties through the choice of different functional compounds, since these are bound covalently to the solid surface under controlled conditions [309, 314, 315].

The most important method of surface modification is grafting, which enables formation of gradient surfaces with the desired wettability, thickness and other physicochemical properties by applying “grafting to” or “grafting from” approaches.

A useful method of preparation of dense arrays of macromolecules with nanometer-scale precision for gradient surfaces is atom transfer radical polymerization (ATRP), which has been used to polymerize, e.g., methacrylates, acrylates, styrenes, vinylpyridines, acrylonitrile and acrylamides [316, 317]. The preparation of a gradient surface using ATRP provides a convenient means for introducing a variety of polymer architectures, including block copolymers, multiarmed stars, hyperbranched polymers and polymer combs with well-defined composition and relatively low molecular weight polydispersity [309, 318, 319].

9.3 Other Techniques

Various methods of biomaterial surface treatment, like vapour-based polymer coating, plasmainduced micropatterning, plasma-based ion implantation, corona discharge, ion sputtering and UV irradiation have been utilized in recent years [309, 310, 320–322]. These techniques are described in detail in recent review articles [305, 309, 310, 320–322].

10 Characterization of Polymer/HAp Nanocomposites for Biomedical Applications

Characterization of nanomaterials at the nanoscale is crucial for understanding and, hence, further promoting increased tissue growth on (nano)materials. Only through the coordination of nanoscale analytical tools with studies that highlight the mechanisms of increased tissue growth in the presence of nanomaterials will design of better implant materials be possible [323]. Since characterization of bulk material differs from an assessment of the surface properties, these issues will be described separately.

10.1 Characterization of Bulk Material

The major physico-chemical methods used for characterization of polymer/HAp composites are microscopic techniques, X-ray diffraction, mechanical testing and thermal analysis methods.

The structure of nanocomposites has typically been established in literature using a wide-angle X-ray diffraction (WAXD) analysis and transmission electron microscope (TEM) observations. Due to its availability and ease of use, WAXD is most commonly used to probe nanocomposite structures. By monitoring the position, shape, and intensity of the basal reflections from the distributed nanoparticles, the nanocomposite structure may be identified [324]. Additional information can be obtained from small-angle X-ray scattering (SAXS) This technique yields quantitative characterization of nanostructure in composite materials [133].

Among the microscopic methods used for characterization of polymer/HAp systems the most important are SEM and TEM. TEM is recommended whenever an in-depth study is required and allows a qualitative understanding of the internal structure, spatial distribution of the various phases, and defect structure of nanocomposites through direct visualization. However, special care must be exercized to ensure that a representative crosssection of the sample is evaluated [325]. 3D TEM is a recent development in the area of high-resolution electron microscopy – it allows visualization of the 3D structure of nanocomposite materials.

Both TEM and WAXD are essential tools for evaluating nanocomposite structure. However, TEM is time-intensive and only gives qualitative information on the sample as a whole, whereas wide-angle peaks in WAXD allow quantification of changes in layer spacing. Typically, when layer spacing exceeds 6–7 nm in intercalated nanocomposites or when the layers become relatively disordered in exfoliated nanocomposites, associated WAXD features weaken to the point of not being useful [133].

The thermal and thermomechanical properties of the polymer/HAp composites (glass transition temperature, melting and crystallization behaviour, thermal stability, crosslinking effects, phase composition, modulus, etc.) can be evaluated by thermal analysis methods, like TG, DSC and DMA. Recently, a modulated temperature DSC (MTDSC) technique has been developed that offers extended temperature profile capabilities by, for example, a sinusoidal wave superimposed on the normal linear temperature ramp [326]. The new capabilities of the MTDSC method in comparison with conventional DSC include separation of reversible and non-reversible thermal events, improved resolution of closely occurring and overlapping transitions, and increased sensitivity of heat capacity measurements [92, 327].

In solid state NMR, the main objective is to connect the measured longitudinal relaxation ($T_1{}^{H}$ s) of protons and ^{13}C nuclei with the quality of dispersion. Since the homogeneity of the dispersion of the nanoparticles within the polymer matrix is of primary importance, this technique is getting increased attention [328] although it is still less commonly used than the other methods.

10.2 Characterization of Surface

Among spectroscopic techniques, FTIR spectroscopy with attenuated total reflection (ATR) is a useful technique for characterization of the surface of polymer/HAp biomaterials. The use of FTIR-ATR is limited by the micrometer range probed by the method, and that the chemical functionalities constituting the gradient are to be clearly different from the substrate. This technique can provide valuable information related to the chemical bonding and specific interactions, like hydrogen bonds, between the composite's components. FTIR spectroscopy makes it possible to determine the phase composition, miscibility of the components and the rate of bone tissue regeneration. For instance, in the work of Blazewicz and Paluszkiewicz, FTIR spectroscopy has been used to analyse the process of bone regeneration taking place in the presence of surfacemodified carbon materials [329].

XPS is another widely employed technique used to characterize the surface of (nano)biomaterials. XPS has a much smaller sampling depth (ca. 10 nm) and is, thus, a more sensitive surface analysis technique than FTIR-ATR. This is because the mean free average path of the X-ray that excites electrons in polymers for XPS is typically only 5–10 nm. XPS has been used extensively in the study of polymer composite surfaces to investigate the extent of surface modification, polymer chain mobility, contamination, and degradation processes.

One technique with even higher sensitivity (sampling depth is 1–2 nm) than XPS is time-of-flight secondary ion mass spectrometry (TOF-SIMS). SIMS is a mass-spectroscopy-based technique for analysing elemental and chemical compositions of the outermost molecular or atomic layer of a solid surface. One important application of SIMS is to study the composition, conformation, orientation and denaturation of proteins on nanomaterial surfaces [323].

Atomic force microscopy (AFM) is a technique that is capable of producing 3D images of polymer nanocomposites samples. In this technique the tip of a probe is rastered over a surface to build up an image of the surface topography. It can achieve very high, even atomic, resolution, of the samples in ambient air or liquid. Another advantage of AFM is that sample preparation is less time-consuming than in scanning or transmission electron microscopy [331].

Microthermal analysis (micro-TA) provides a characterization tool capable of imaging samples in a variety of modes, including those of current AFM technology, with very high resolution. It is also capable of simultaneous characterization of the properties of the sample in a spatially resolved way using thermal analysis. Scanning thermal imaging (SThM) is a characterization technique combining thermal analysis and AFM, in which the normal probe has been replaced by one with a resistive heater at the tip. This is used to build up an image based on the contrast in apparent thermal conductivity, as well as providing the usual topographic information via the optical lever feedback circuit. Chemical analysis is then also possible using localized desorption pyrolysis gas chromatography mass spectrometry (or just mass spectrometry) and highresolution localized IR spectrometry [330].

Other techniques that are used for surface characterization of polymer/HAp nanomaterials include ellipsometry which enables measurement of the thicknesses and composition of the layers in a multilayer thin film, especially if the experiment is performed over a range of wavelengths of the incident beam or a range of angles of incidence. In addition, ellipsometry is frequently applied to characterization of the thickness of proteins adsorbed on gradient surfaces [309].

Because wetting determines the contact angle, measurement of the latter is one of the most sensitive techniques for studying a surface. Contact angles on gradient surfaces can be measured using standard liquids with known free surface energy and goniometric or tensiometric equipment.

11 Nanomaterial–Tissue Interactions

Nanomaterial–tissue interactions are currently the subject of intensive investigations because they strongly influence properties and thus extend or limit new application areas. Nanomaterials, due to high surface areas and roughness, have large surface energy and unique surface properties arising from surface defects, nonregular grain boundaries, etc. All of these unusual properties are inherent for nanomaterials and will affect interactions with proteins, which are natural nanoscale entities. For example, the increased surface area and nanoscale surface features of nanomaterials can provide more available sites for protein adsorption and, thus, alter the capability for cellular interactions. One straightforward explanation lies in the fact that natural tissues and associated extracellular matrices are composed of nanostructured materials [323].

It is not yet known exactly how nanoparticles interact with tissue. It is suggested that nanoparticles, because of their small size, may act like haptens to modify protein structures, either by altering their function or rendering them antigenic, thus raising their potential for autoimmune effects [332]. When the particles fall within a range of about 20–50 nm, they are of a size that can enter the central nervous system and cells. In addition, macrophages patrolling on the alveolar surfaces of the lung appear to have difficulty in recognizing particles of less than 70 nm as being "foreign", thus allowing them to gain access to the pulmonary interstitium, a potentially vulnerable anatomical compartment [333]. The large surface area of nanoparticles could also mean that they can absorb proteins. Adsorption of different proteins occuring with different nanoparticle surfaces could also modify how nanoparticles are handled by macrophages and other cells and might make nanoparticles potent at activating the complement system. Not only could adverse effects be induced, but also the interactions of nanoparticles with cell and subcellular structures and their distribution/biokinetics are likely be very different from those of larger-sized particles [334] Risk of bioaccumulation and redistribution from their site of disposition in organs and tissues has to be considered.

12 Potential Hazards of Nanoparticles

The same unique physical and chemical properties that make nanomaterials so attractive could be associated with their potentially calamitous effects on cells and tissues. There is emerging concern that nanosized particles merit a more rigorous assessment of their potential effects on health (and the associated control requirements) because their surface area impact and toxicity might be significantly greater than those of larger particles. However, specific mechanisms and pathways through which nanomaterials exert their toxic effects remain unknown. Although several laboratories have reported potential toxic effects of nanoparticles on different types of cells in vitro, there are only a few publications on in vivo toxicity of nanomaterials [335, 336].

The properties of nanomaterials depend on a wider range of parameters than macroscopic materials across many length scales. Relationships between the properties of nanomaterials and their size, composition, impurities (both internally and superficially), the surface chemistry and degree of agglomeration are not well established. As an example one can present the relationship between size of particle and phase stability of titanium dioxide. Macroscopically the rutile phase is stable and the anatase phase metastable, but when the particle size is under $\sim$20 nm this situation is reversed. There is an intrinsic relationship between the two dependencies, size and phase [337].

Although the small size of particles is what makes nanotechnology so useful in medicine and materials engineering, it is also one of the main factors that might make them potentially dangerous to human health. The general opinion is that the smaller the particles, the more reactive and toxic are their effects. This is because any intrinsic properties of particles will probably be emphasized with the increase in surface energy per unit mass. Studies have shown that ultrafine particles (e.g., fine scale carbon graphite), irrespective of their chemical composition, are potent inducers of inflammatory lung injury. It is highly unlikely that all the materials used in construction of nanoparticles will be biologically inert and many quantum dot core metals are already known to be toxic to vertebrate systems at relatively low concentrations (ppm) [333].

13 Future Trends

The future of new nanostructured biomedical materials is dependent on the development of an enhanced knowledge base and greater understanding of (a) the structure–property relationships, and (b) the molecular, cellular and tissue interactions between materials. Nature provides a multiplicity of materials, architectures, systems and functions – the latter may be different for a unique component operating in the same organism. For instance, a protein such as CollI presents different morphologies in tissues that have to perform different functions. Associated with HAp crystals CollI creates elastic (suitable Young's modulus) and shock-resistant tissues in bone; it behaves like an elastomer with low rigidity and high reversible deformation to rupture in tendons, or posseses the optical properties of transparency in the cornea. Well-designed and appropriately characterized composite materials, adaptive materials, hierarchically structured nanomaterials, 3D networks, hybrids with enhanced properties and so on, should thus become a major preoccupation of research in the field of advanced technologies for regenerative medicine [20, 338].

The development of biomimetic and bioinspired multifunctional composite materials will have a significant influence on the development of novel, alternative methods and approaches for treatment of bone injuries and for surgical techniques in bone diseases, and could improve the effectiveness of the treatment of some diseases. In addition to these potential clinical applications, such engineered biomaterials can contribute to fundamental science, including better understanding of the interactions of materials with living cell systems.

Creation of the next generation of biomaterials with well-defined nanotopography that can elicit the desired cellular and tissue response will be the primary task for the next decade. Biomaterials proposed as optimum scaffolds for cell growth should be biocompatible, osteoinductive and/or osteoconductive, integrative, porous, and mechanically compatible with native bone to fulfil their desired role as bone implants and substitutes. Nanotechnology can provide an alternative platform with higher mechanical strength, enhanced functionality and resorbability for improving the quality of bone tissue regeneration materials [339].

Polymer nanocomposites generally offer new possibilities as materials for bone tissue regeneration because the sizes of the nanoparticles are similar to those of bone components and they may interact at the nanolevel. Specifically, polymer/HAp nanocomposites are promising materials for future applications because HAp is osteoinductive and its nanosized structure resembles that of bone. The limited commercial supply of nanoHAp and the difficulty of uniformly dispersing nanoscale particles in a viscous polymer matrix using the currently available methods hinder the systematic examination of the polymer/nanoHAp structure and properties and of the mechanisms of cell–tissue interactions. Hence, one can assume that new research directions will focus on the systematic determination and optimization of the processing–structure–property relationships in polymer/HAp nanomaterials using high-resolution imaging and analysis tools and the subsequent evaluation of the long-term consequences of nanoHAp biomedical uses in terms of its degradation behaviour, toxicity and tissue response.

14 Conclusions

Numerous novel polymer/(nano)HAp materials for bone tissue re-generation have been developed in recent years, as illustated in this review. Nondegradable or degradable polymeric matrices offer a wide range of physical and mechanical properties, and they display, depending on the polymer type, different biodegradation characteristics. Since most polymers are not compatible with bone tissue, an appropriate modification of their structure and properties by incorporation of ceramics is necessary to obtain materials that mimic the structural and morphological organization of bone. One of the most important ceramic materials is HAp which, as the closest synthetic equivalent to human bone mineral and having excellent biocompatibility and osteoconductivity, is a suitable candidate for bone tissue regeneration. HAp can facilitate the formation of new bone without resorption and can interact with the living system. It has been found that if HAp occurs in its nanostructured form with a very high surface area, the interactions with the lowest hierarchical levels of bone's structure (i.e., molecular structure of apatite, collagen and non-collagenous organic proteins) become predominant. However, the mechanisms of polymer/nanoHAp interactions with bone tissue still remain to be discovered. HAp nanoparticles may act like haptens to modify the protein structures or they may absorb them. Their distribution and bio-kinetics are also likely to be quite different from those of larger-sized particles and, therefore, the risks of bioaccumulation and redistribution have to be considered. Since the relationships between the properties of nanoHAp to its particle size, composition (including the effect of impurities) and surface chemistry are not yet established, special care must be taken when testing and using novel biomimetic polymer/HAp systems. Nanostructured biomedical polymer/HAp materials offer a wide range of properties that can be utilized for various applications in bone tissue engineering, but their future application does depend on the development of an enhanced knowledge base and understanding of the processing–structure–property relationships and the complex interactions with human tissue. New research strategies will probably focus on optimization of the polymer/HAp processing conditions in order to control the interfacial bonding and strength, as well as on the further understanding of the mechanisms of nanocomposite–tissue in vivo interactions required to promote effective bone regeneration.

Acknowledgements One of the authors (KP) is grateful to the Polish Ministry of Science and Higher Education for financial support under contract No. PBZ/MEiN/01/2006/15.

References

1. Seal BL, Otero TC, Panitch A (2001) Polymeric biomaterials for tissue and organ regeneration. Mater Sci Eng 34:147–230
2. Hall B (2005) Bones and cartilage: developmental skeletal biology. Academic, San Diego
3. Russias J, Saiz E, Nalla RK et al (2006) Microspheres as building blocks for hydroxyapatite/polylactide. J Mater Sci 41:5127–5133
4. Rodrigues CVM, Serricella P, Linhares ABR et al (2003) Characterization of a bovine collagen–hydroxyapatite composite scaffold for bone tissue engineering. Biomaterials 24:4987–4997
5. Tan KH, Chua CK, Leong KF et al (2003) Scaffold development using selective laser sintering of polyetheretherketone–hydroxyapatite biocomposite blends. Biomaterials 24:3115–3123
6. Teoh SH (2004) Engineering materials for biomedical applications.World Scientific, Singapore
7. Liang W, Rahaman MN, Day DE et al (2008) Bioactive borate glass scaffold for bone tissue engineering. J Non-Crystal Solids 354:1690–1696
8. Wang M (2003) Developing bioactive composite materials for tissue replacement. Biomaterials 24:2133–2151
9. Bouyer E, Gitzhofer F, Boulos MI (2000) Morphological study of hydroxyapatite nanocrystal suspension. J Mater Sci Mater Med 11:523–531
10. Roeder RK, Converse GL, Kane RJ et al (2008) Hydroxyapatite-reinforced polymer biocomposites for synthetic bone substitutes. Biol Mater Sci 3:38–45
11. Converse GL, Yue W, Roeder RK (2007) Processing and tensile properties of hydroxyapatite-whisker-reinforced polyetheretherketone. Biomaterials 28:927–935
12. Rho JY, Kuhn-Spearing L, Zioupos P (1998) Mechanical properties and the hierarchical structure of bone. Med Eng Phys 20:92–102
13. Fratzl P, Weinkamer R (2007) Nature's hierarchical materials. Prog Mater Sci 52:1263–1334
14. Knott L, Bailey AJ (1998) Collagen cross-links in mineralizing tissues: a review of their chemistry, function, and clinical relevance. Bone 22:181–187
15. Currey JD (1999) The design of mineralised hard tissues for their mechanical functions. J Exp Biol 202:3285–3294

16. Currey JD (2003) How well are bones designed to resist fracture? J Bone Min Res 18: 591–598
17. Gupta HS, Seto J, Wagermaier W et al (2006) Cooperative deformation of mineral and collagen in bone at the nanoscale. Proc Natl Acad Sci USA 103:17741–17746
18. Gupta HS, Stachewicz U, Wagermaier W et al (2006) Mechanical modulation at the lamellar level in osteonal bone. J Mater Res 21:1913–1921
19. Tai K, Dao M, Suresh S et al (2007) Nanoscale heterogeneity promotes energy dissipation in bone. Nature Mater 6:454–462
20. Tamerler C, Sarikaya M (2007) Molecular biomimetics: utilizing nature's molecular ways in practical engineering. Acta Biomater 3:289–299
21. Olszta MJ, Cheng X, Jee SS et al (2007) Bone structure and formation: a new perspective. Mater Sci Eng R58:77–116
22. Meyers MA, Chen PY, Lin AYM et al (2008) Biological materials: structure and mechanical properties. Progr Mater Sci 53:1–206
23. Murugan R, Ramakrishna S (2005) Development of nanocomposites for bone grafting. Compos Sci Techn 65:2385–2406
24. Isenberg BC, Wong JY (2006) Building structure into engineered tissues. Mater Today 9:54–60
25. Peters F, Schwartz K, Epple M (2000) The structure of bone studied with synchrotron X-ray diffraction, X-ray absorption spectroscopy and thermal analysis. Thermochim Acta 361: 131–138
26. Murugan R, Ramakrishna S (2005) Aqueous mediated synthesis of bioresorbable nanocrystalline hydroxyapatite. J Cryst Growth 274:209–213
27. Murugan R, Ramakrishna S (2006) Production of ultra-fine bioresorbable carbonated hydroxyapatite. Acta Biomater 2:201–206
28. Nihouannen DL, Saffarzadeh A, Gauthier O et al (2008) Bone tissue formation in sheep muscles induced by a biphasic calcium phosphate ceramic and fibrin glue composite. J Mater Sci Mater Med 19:667–675
29. Siddharthan A, Seshadri SK, Sampath Kumar TS (2004) Microwave accelerated synthesis of nanosized calcium deficient hydroxyapatite. J Mater Sci Mater Med 15:1279–1284
30. Hao J, Liu Y, Zhou S et al (2003) Investigation of nanocomposites based on semi-interpenetrating network of [L-poly(ε-caprolactone)]/[net-poly(ε-caprolactone)] and hydroxyapatite nanocrystals. Biomaterials 24:1531–1539
31. Deng X, Hao JY, Wang C (2001) Preparation and mechanical properties of nanocomposites of poly(D,L-lactide) with Ca-deffcient hydroxyapatite nanocrystals. Biomaterials 22:2867–2873
32. Jokanovic V, Izvonar D, Dramicanin MD et al (2006) Hydrothermal synthesis and nanostructure of carbonated calcium hydroxyapatite. J Mater Sci Mater Med 17:539–546
33. Nayar S, Sinha MK, Basu D et al (2006) Synthesis and sintering of biomimetic hydroxyapatite nanoparticles for biomedical applications. J Mater Sci Mater Med 17:1063–1068
34. Patel N, Gibson IR, Ke S et al (2001) Calcining influence on the powder properties of hydroxyapatite. J Mater Sci Mater Med 12:181–188
35. Slosarczyk A, Paszkiewicz Z, Paluszkiewicz C (2005) FTIR and XRD evaluation of carbonated hydroxyapatite powders synthesized by wet methods. J Molecul Struct 744–747:657–661
36. Rapacz-Kmita A, Paluszkiewicz Cz, Slosarczyk A et al (2005) FTIR and XRD investigations on the thermal stability of hydroxyapatite during hot pressing and pressureless sintering processes. J Molecul Struct 744–747:653–656
37. Zhang Y, Zhou L, Xue N et al (2003) Oriented nano-structured hydroxyapatite from the template. Chem Phys Lett 376:493–497
38. Zhou WY, Wang M, Cheung WL et al (2008) Synthesis of carbonated hydroxyapatite nanospheres through nanoemulsion. J Mater Sci Mater Med 19:103–110
39. Solans C, Izquierdo P, Nolla J et al (2005) Nano-emulsions. Curr Opin Coll Interface Sci 10:102–110
40. Landi E, Logroscino G, Proietti L et al (2008) Biomimetic Mg-substituted hydroxyapatite: from synthesis to in vivo behaviour. J Mater Sci Mater Med 19:239–247

41. Haberko K, Bucko M, Brzezinska-Miecznik B et al (2006) Natural hydroxyapatite–its behaviour during heat treatment. J Eur Cer Soc 26:537–542
42. Gao Y, Cao WL, Wang XY et al (2006) Characterization and osteoblast-like cell compatibility of porous scaffolds: bovine hydroxyapatite and novel hydroxyapatite artificial bone. J Mater Sci Mater Med 17:815–823
43. Ruksudjarit A, Pengpat K, Rujijanagul G et al (2008) Synthesis and characterization of nanocrystalline hydroxyapatite from natural bovine bone. Curr Appl Phys 8:270–272
44. Ono I, Tateshita T, Nakajima T (2000) Evaluation of a high density polyethylene fixing system for hydroxyapatite ceramic implants. Biomaterials 21:143–151
45. Yamaguchi I, Tokuchi K, Fukuzaki H et al (2001) Preparation and microstructure analysis of chitosan/hydroxyapatite nanocomposites. J Biomed Mater Res 55:20–27
46. Bonfield W, Grynpas MD, Tully AE et al (1981) Hydroxyapatite reinforced polyethylene – a mechanically compatible implant material for bone replacement. Biomaterials 2:185–186
47. Fan JP, Tsui CP, Tang CY et al (2004) Influence of interphase layer on the overall elasto-plastic behaviors of HA/PEEK biocomposite. Biomaterials 25:5363–5373
48. Liu X, Ma PX (2004) Polymeric scaffolds for bone tissue engineering. Ann Biomed Eng 32:477–486
49. Yaszemski MJ et al (1996) The evolution of bone transplantation: molecular, cellular, and tissue strategies to engineer human bone. Biomaterials 7:175–185
50. Crane GM, Ishaug SL, Mikos AG (1995) Bone tissue engineering. Nature Med 1:1322–1324
51. Chapekar MS (2000) Tissue engineering: challenges and opportunities. J Biomed Mater Res 53:617–20
52. Kenny SM, Buggy M (2003). Bone cements and fillers: a review. J Mater Sci Mater Med 14:923–938
53. Middleton JC, Tipton AJ (2000) Synthetic biodegradable polymers as orthopedic devices. Biomaterials 21:2335–2346
54. Burg KJL, Porter S, Kellam JF (2000) Biomaterial developments for bone tissue engineering. Biomaterials 21:2347–2359
55. Eschbach L (2000) Nonresorbable polymers in bone surgery. Injury, Int J Care Injured 31: S-D22–27
56. Mano JF, Sousa RA, Boesel LF et al (2004) Bioinert, biodegradable and injectable polymeric matrix composites for hard tissue replacement: state of the art and recent developments. Compos Sci Techn 64:789–817
57. Blazewicz S (2006) Non-metalic multifunctional composites in biomaterials engineering. In: Nadolny AJ (ed) Biomaterials in regenerative medicine. Polish Academy of Sciences, Warszawa
58. Hutmacher DW (2000) Scaffolds in tissue engineering bone and cartilage. Biomaterials 21:2529–2543
59. Vert M, Li MS, Spenlehauer G et al (1992) Bioresorbability and biocompatibility of aliphatic polyesters. J Mater Sci 3:432–446
60. Nair LS, Laurencin CT (2007) Biodegradable polymers as biomaterials. Prog Polym Sci 32:762–798
61. Goepferich A (1997) Polymer bulk erosion. Macromolecules 30:2598–2604
62. Goepferich A, Tessmar J (2002) Polyanhydride degradation and erosion. Adv Drug Deliv Rev 54:911–931
63. Gopferich A (1996) Mechanisms of polymer degradation and erosion. Biomaterials 17: 103–114
64. Katti DS, Lakshmi S, Langer R et al (2002). Toxicity, biodegradation and elimination of polyanhydrides. Adv Drug Deliv Rev 54:933–961
65. Temenof JS, Mikos GA (2000) Injectable biodegradable materials for orthopedic tissue engineering. Biomaterials 21:2405–2412
66. Chandra R, Rustgi R (1998) Biodegradable polymers. Prog Polym Sci 23:1273–1335
67. Roether JA, Boccaccini AR, Hench LL et al (2002) Development and in vitro characterisation of novel bioresorbable and bioactive composite materials based on polylactide foams and Bioglasss for tissue engineering applications. Biomaterials 23:3871–3878

68. Stoch L, Blazewicz S (eds) (2000) Biomaterialy. Biocybernetyka i inzynieria biomedyczna, vol. 4. Polish Academy of Science, Warszawa
69. Lee HJ, Kim, SE, Choi HW et al (2007) The effect of surface-modified nano-hydroxyapatite on biocompatibility of poly(ε-caprolactone)/hydroxyapatite nanocomposites. Eur Polym J 43:1602–1608
70. Mark HF (2004) Encyclopedia of polymer science and technology. Wiley, New York
71. Wang M, Bonfield W (2001) Chemically coupled hydroxyapatite-polyethylene composites: structure and properties. Biomaterials 22:1311–1320
72. Wang M, Joseph R, Bonfield W (1998) Hydroxyapatite-polyethylene composites for bone substitution: effects of ceramic particle size and morphology. Biomaterials 19:2357–2366
73. Nazhat SN, Joseph R, Wang M et al (2000) Dynamic mechanical characterization of hydroxyapatite reinforced polyethylene: effect of particle size. J Mater Sci Mater Med 11:621–628
74. Wang M, Chandrasekaran M, Bonfield W (2002) Friction and wear of hydroxyapatite reinforced high density polyethylene against stainless steel counterface. J Mater Sci Mater Med 13:607–611
75. Wang M, Porter D, Bonfield W (1994) Processing, characterisation, and evaluation of hydroxyapatite reinforced polyethylene composites. Br Ceram Trans 93:91–95
76. Sousa RA, Reis RL, Cunha AN et al (2003) Coupling of HDPE/hydroxyapatite composites by silane-based methodologies. J Mater Sci Mater Med 14:475–487
77. Joseph R, McGregor WJ, Martyn MT et al (2002) Effect of hydroxyapatite morphology/surface area on the rheology and processability of hydroxyapatite filled polyethylene composites. Biomaterials 23:4295–4302
78. Reis RL, Cunha AM, Oliveira MJ et al (2001) Relationship between processing and mechanical properties of injection molded high molecular mass polyethylene + hydroxyapatite composites. Mat Res Innovat 4:263–272
79. Roeder RK, Sproul MM, Turner CH (2003) Hydroxyapatite whiskers provide improved mechanical properties in reinforced polymer composites. J Biomed Mater Res 67A:801–812
80. Yari Sadi A, Shokrgozar MA, Homaeigohar S et al (2006) The effect of partially stabilized zirconia on the biological properties of HA/HDPE composites in vitro. J Mater Sci Mater Med 17:407–412
81. Fang LM, Leng Y, Gao P (2005) Processing of hydroxyapatite reinforced ultrahigh molecular weight polyethylene for biomedical applications. Biomaterials 26:3471–3478
82. Fang L, Leng Y, Gao P (2006) Processing and mechanical properties of HA/UHMWPE nanocomposites. Biomaterials 27:3701–3707
83. Fang L, Gao P, Leng Y (2007) High strength and bioactive hydroxyapatite nano-particles reinforced ultrahigh molecular weight polyethylene. Composites Part B 38:345–351
84. Smolko E, Romero G (2007) Studies on crosslinked hydroxyapatite-polyethylene composite as a bone-analogue material. Rad Phys Chem 76:1414–1418
85. Albano C, Cataño L, Figuera L et al (2009) Evaluation of a composite based on high-density polyethylene filled with surface-treated hydroxyapatite, Polym Bull 62:45–55
86. Huang M, Feng J, Wang J et al (2003) Synthesis and characterization of nano-HA/PA66 composites. J Mater Sci Mater Med 14:655–660
87. Wei J, Li Y, Chen W et al (2003) A study on nano-composite of hydroxyapatite and polyamide. J Mater Sci 38:3303–3306
88. Wei J, Li Y (2004) Tissue engineering scaffold material of nano-apatite crystals and polyamide composite. Eur Polym J 40:509–515
89. Wei J, Li Y, Lau KT (2007) Preparation and characterization of a nanoapatite/polyamide6 bioactive composite. Composites Part B 38:301–305
90. Zhang X, Li Y, Lv G et al (2006) Thermal and crystallization studies of nano-hydroxyapatite reinforced polyamide 66 biocomposites. Polym Degrad Stab 91:1202–1207
91. Masamoto J (1993) Modern polyacetals. Prog Polym Sci 18:1–84
92. Pielichowski K, Flejtuch K (2004) Some comments on the melting and recrystallization of polyoxymethylene by high-speed and StepScan differential scanning calorimetry. Polimery 49:80–82

93. Brydson JA (1993) Polyacetals and related materials. In: Plastic materials. Butterworth Heinemann, Oxford, pp. 531–555
94. Thompson MS, Northmore-Ball MD, Tanner KE (2001) Tensile mechanical properties of polyacetal after one and six month's immersion in ringer's solution. J Mater Sci Mater Med 12:883–887
95. Thompson MS (2001) The design of a novel hip resurfacing prosthesis. PhD thesis, London
96. Moore DJ, Freeman MAR, Revell PA et al (1998) Can a total knee replacement prosthesis be made entirely of polymers? J Arthroplasty 13:388–395
97. McKellop HA, Milligan HL, Röstlund T (1996) Long term biostability of polyacetal (delrin) implants. J Heart Valve Disease 5:S238–S242
98. Zilberman U (2005) Formaldehyde from POM brackets. Am J Orthodont Dentofac Orthop 128:147–148
99. Pielichowska K (2008) Preparation of polyoxymethylene/hydroxyapatite nanocomposites by melt processing. Int J Mater Form 1:941–944
100. Pielichowska K (2008) Bioactive POM/HAp nanocomposites for long term bone implants. Mod Polym Mater Env Appl 3:179–184
101. Pielichowska K (2009) unpublished data
102. Wang M, Yue CY, Chua B (2001) Production and evaluation of hydroxyapatite reinforced polysulfone for tissue replacement. J Mater Sci Mater Med 12:821–826
103. Spanos N, Deimede V, Koutsoukos PG (2002) Functionalization of synthetic polymers for potential use as biomaterials: selective growth of hydroxyapatite on sulphonated polysulphone. Biomaterials 23:947–953
104. Szaraniec B, Rosol P, Chlopek J (2005) Carbon composite material and polysulfone modified by nano-hydroxyapatite. e-Polymers 30:1–7
105. Chlopek J, Rosol P, Morawska-Chochol A (2006) Durability of polymer-ceramics composite implants determined in creep tests. Compos Sci Techn 66:1615–1622
106. Abu Bakar MS, Cheng MHW, Tang SM et al (2003) Tensile properties, tension–tension fatigue and biological response of polyetheretherketone–hydroxyapatite composites for load-bearing orthopedic implants. Biomaterials 24:2245–2250
107. Yu S, Hariram KP, Kumar R et al (2005) In vitro apatite formation and its growth kinetics on hydroxyapatite/polyetheretherketone biocomposites. Biomaterials 26:2343–2352
108. Abu Bakar MS, Cheang P, Khor KA (2003) Mechanical properties of injection molded hydroxyapatite-polyetheretherketone biocomposites. Compos Sci Techn 63:421–425
109. Abu Bakar MS (2003) Tensile properties and microstructural analysis of spheroidized hydroxyapatite/poly(etheretherketone) biocomposites. Mater Sci Eng A345:55–63
110. Meenan BJ, McClorey C, Akay M (2000) Thermal analysis studies of poly(etheretherketone)/ hydroxyapatite biocomposite mixtures. J Mater Sci Mater Med 11:481–489
111. Pielichowski K, Flejtuch K, Pielichowski J (2004) Step-scan alternating DSC study of melting and crystallisation in poly(ethylene oxide). Polymer 45:1235–1242
112. Dalby MJ, Silvio LD, Harper EJ et al (2002) Increasing hydroxyapatite incorporation into poly(methylmethacrylate) cement increases osteoblast adhesion and response. Biomaterials 23:569–576
113. Vallo CI, Montemartini PE, Fanovich MA et al (1999) Polymethylmethacrylate-based bone cement modified with hydroxyapatite. J Biomed Mater Res (Appl Biomater) 48:150–158
114. Cunin G, Boissonnet H, Petite H et al (2000) Experimental vertebroplasty using osteoconductive granular material. Spine 25:1070–1076
115. Kim YS, Kang YH, Kim JK (1994) The effect of bone mineral particles on the porosity of bone cement. Biomed Mater Eng 4:37–46
116. Lu JX, Huang ZW, Tropiano P (2002) Human biological reactions at the interface between bone tissue and polymethylmethacrylate cement. J Mater Sci Mater Med 13:803–809
117. Barralet JE, Gaunt T, Wright AJ et al (2002) Effect of porosity by compaction on compressive strength and microstructure of calcium phosphate cement. J Biomed Mater Res (Appl Biomater) 63:1–9
118. Kim SB, Kim YJ, Yoon TL et al (2004) The characteristics of a hydroxyapatite–chitosan–PMMA bone cement. Biomaterials 25:5715–5723

119. De Santis R, Ambrogi V, Carfagna C et al (2006) Effect of microencapsulated phase change materials on the thermo-mechanical properties of poly(methyl-methacrylate) based biomaterials. J Mater Sci Mater Med 17:1219–1226
120. Pielichowski K, Flejtuch K (2002) Differential scanning calorimetry studies on polyethylene glycol with different molecular weights for thermal energy storage materials. Polym Adv Techn 13:690–696
121. Migliaresi C, Fambri L, Kolarik J (1994) Polymerization kinetics, glass transition temperature and creep of acrylic bone cements. Biomaterials 15:875–881
122. Liu Q, de Wijn JR, Blitterswijk CA (1998) Covalent bonding of PMMA, PBMA, and poly(HEMA) to hydroxyapatite particles. J Biomed Mater Res;40:257–263.
123. Kwon SW, Kim YS, Woo YK et al (1997) Hydroxyapatite impregnated bone cement: in vitro and in vivo studies. Biomed Mater Eng 7:129–40
124. Harper EJ, Behiri JC, Bonfield W (1995) Flexural and fatigue properties of a bone cement based upon polyethylmethacrylate and hydroxyapatite. J Mater Sci Mater Med 6:799–803
125. Moursi AM, Winnard AV, Winnard PL et al (2002) Enhanced osteoblast response to a polymethylmethacrylate–hydroxyapatite composite. Biomaterials 23:133–144
126. Skrtic D, Stansbury JW, Antonucci JM (2003) Volumetric contraction and methacrylate conversion in photopolymerized amorphous calcium phosphate/methacrylate composites. Biomaterials 24:2443–2449
127. Vert M (2007) Polymeric biomaterials: Strategies of the past vs. strategies of the future. Prog Polym Sci 32:755–761
128. Ratner BD, Hoffman AS, Schoen FJ et al (1996) Biomaterials science. In: Lemons JE (ed) An introduction to materials in medicine. Academic, San Diego
129. Gilding DK, Reed AM (1979) Biodegradable polymers for use in surgery – polyglycolic/poly(lactic acid) homo- and copolymers. Polymer 20:1459–1464
130. Barrows TH (1986) Degradable implant materials: a review of synthetic absorbable polymers and their applications. Clin Mater 1:233–257
131. Gunatillake P, Mayadunne R, Adhikari R (2006) Recent developments in biodegradable synthetic polymers. Biotechnol Ann Rev 12:301–347
132. Li S (1999) Hydrolytic degradation characteristics of aliphatic polyesters derived from lactic and glycolic acids. J Biomed Mater Res 48:342–353
133. Ray SS, Bousmina M (2005) Biodegradable polymers and their layered silicate nanocomposites: in greening the 21st century materials world. Prog Mater Sci 50:962–1079
134. Bergsma JE, de Bruijn WC, Rozema FR et al (1995) Late degradation tissue response to poly(L-lactide) bone plates and screws. Biomaterials 16:25–31
135. Ignjatovic N, Tomic S, Dakic M et al (1999) Synthesis and properties of hydroxyapatite/poly-L-lactide composite biomaterials. Biomaterials 20:809–816
136. Ignjatovic N, Savic H, Najman S et al (2001) A study of HAp/PLLA composite as a substitute for bone powder, using FT-IR spectroscopy. Biomaterials 22:571–575
137. Ignjatovic N, Uskokovic D (2004) Synthesis and application of hydroxyapatite/polylactide composite biomaterial. Appl Surface Sci 238:314–319
138. Jin S, Gonsalves KE (1999) Functionalized copolymers and their composites with polylactide and hydroxyapatite. J Mater Sci Mater Med 10:363–368
139. Ural E, Kesenci K, Fambri L et al (2001) Poly(D,L-cactide/ε-caprolactone)/hydroxyapatite composites. Biomaterials 21:2147–2154
140. Durucan C, Brown PW (2000) Low temperature formation of calcium-deficient hydroxyapatite-PLA/PLGA composites. J Biomed Mater Res 51:717–725
141. Furukawa T, Matsusue Y, Yasunaga T et al (2000) Biodegradation behavior of ultra-high-strength hydroxyapatite/poly(L-lactide) composite rods for internal fixation of bone fractures. Biomaterials 21:889–898
142. Hasegawa S, Ishii S, Tamura J et al (2006) A 5–7 year in vivo study of high-strength hydroxyapatite/poly(L-lactide) composite rods for the internal fixation of bone fractures. Biomaterials 27:1327–1332

143. Shikinami Y, Okuno M (1999) Bioresorbable devices made of forged composites of hydroxyapatite (HA) particles and poly-L-lactide (PLLA): Part I. Basic characteristics. Biomaterials 20:859–877
144. Shikinami Y, Okuno M (2001) Bioresorbable devices made of forged composites of hydroxyapatite (HA) particles and poly L-lactide (PLLA). Part II: practical properties of miniscrews and miniplates. Biomaterials 22:3197–3211
145. Shikinami Y, Matsusue Y, Nakamura T (2005) The complete process of bioresorption and bone replacement using devices made of forged composites of raw hydroxyapatite particles/poly L-lactide (F-u-HA/PLLA). Biomaterials 26:5542–5551
146. Hong Z, Qiu X, Sun J et al (2004) Grafting polymerization of L-lactide on the surface of hydroxyapatite nano-crystals. Polymer 45:6699–6706
147. Hong Z, Zhang P, He C et al (2005) Nano-composite of poly(L-lactide) and surface grafted hydroxyapatite: mechanical properties and biocompatibility. Biomaterials 26:6296–6304
148. Mathieu LM, Mueller TL, Bourban PB et al (2006) Architecture and properties of anisotropic polymer composite scaffolds for bone tissue engineering. Biomaterials 27:905–916
149. Xu X, Chen X, Liu A et al (2007) Electrospun poly(L-lactide)-grafted hydroxyapatite/poly (L-lactide) nanocomposite fibers. Eur Polym J 43:3187–3196
150. Zheng X, Zhou S, Li X et al (2006) Shape memory properties of poly(D,L-lactide)/ hydroxyapatite composites. Biomaterials 27:4288–4295
151. Gay S, Arostegui S, Lemaitre J (2009) Preparation and characterization of dense nanohydroxyapatite/PLLA composites. Mater Sci Eng C29:172–177
152. Lewis OG, Fabisial W (1997) Sutures. In: Kirk-Othmer encyclopedia of chemical technology, 4th edn. Wiley, New York
153. Thomson RC, Yaszemski MJ, Powers JM et al (1995) Fabrication of biodegradable polymer scaffolds to engineer trabecular bone. J Biomater Sci Polym Ed 7:23–28
154. Thomson R (1998) Hydroxyapatite fiber reinforced poly(a-hydroxy ester) foams for bone regeneration. Biomaterials 19:1935–1943
155. Buczynska J, Pamula E, Blażewicz S et al (2006) Modification of resorbable fibres with hydroxyapatite. Eng Biomater 58–60:85–89
156. Rosol P, Chlopek J, Pielichowska K et al (2005) The effect of modifying phases on properties of poli(lactido-co-glycolide) at in vitro conditions. Eng Biomater 47–53:88–92
157. Chlopek J, Morawska-Chochol A, Rosol P (2006) The mechanism of bone tissue regeneration after implantation of resorbable polymer composite modified with hydroxyapatite. Eng Biomater 58–60:98–101
158. Petricca SE, Marra KG, Kumta PN (2006) Chemical synthesis of poly(lactic-co-glycolic acid)/hydroxyapatite composites for orthopaedic applications. Acta Biomater 2:277–286
159. Kim SS, Ahn KM, Park MS et al (2007) A poly(lactide-co-glycolide)/hydroxyapatite composite scaffold with enhanced osteoconductivity. J Biomed Mater Res 80A:206–215
160. Zhang P, Hong Z, Yu T et al (2009) In vivo mineralization and osteogenesis of nanocomposite scaffold of poly(lactide-co-glycolide) and hydroxyapatite surface-grafted with poly (L-lactide). Biomaterials 30:58–70
161. Kesel CD, Wauven CV, David C (1997) Biodegradation of polycaprolactone and its blends with poly(vinylalcohol) by micro-organisms from a compost of house-hold refuse. Polym Degrad Stab 55:107–113
162. Ishiaku US, Pang KW, Lee WS et al (2002) Mechanical properties and enzymic degradation of thermoplastic and granular sago starch filled poly(ε-caprolactone). Eur Polym J 38:393–401
163. Ciapetti G, Ambrosio L, Savarino L et al (2003) Osteoblast growth and function in porous poly ε-caprolactone matrices for bone repair: a preliminary study. Biomaterials 24:3815–3824
164. Causa F, Netti PA, Ambrosio L et al (2006) Poly-ε-caprolactone/hydroxyapatite composites for bone regeneration: in vitro characterization and human osteoblast response. J Biomed Mater Res 76A:151–162
165. Azevedo MC, Reis RL, Claase MB et al (2003) Development and properties of polycaprolactone/hydroxyapatite composite biomaterials. J Mater Sci Mater Med 14:103–107

166. Zinn M, Witholt B, Egli T (2001) Occurrence, synthesis and medical application of bacterial polyhydroxyalkanoate. Adv Drug Deliv Rev 53:5–21
167. Lugg H, Sammons RL, Marquis PM et al (2008) Polyhydroxybutyrate accumulation by a Serratia sp. Biotechnol Lett 30:481–491
168. Freier T (2006) Biopolyesters in tissue engineering applications. Adv Polym Sci 203:1–61
169. Pouton CW, Akhtar S (1996) Biosynthetic polyhydroxyalkanoates and their potential in drug delivery. Adv Drug Deliv Rev 18:133–162
170. Chen C, Dong L, Yu PH (2006) Characterization and properties of biodegradable poly(hydroxyalkanoates) and 4,4-dihydroxydiphenylpropane blends: intermolecular hydrogen bonds, miscibility and crystallization. Eur Polym J 42:2838–2848
171. Chen LJ, Wang M (2002) Production and evaluation of biodegradable composites based on PHB–PHV copolymer. Biomaterials 23:2631–2639
172. Zhao K, Deng Y, Chen J Ch et al (2003) Polyhydroxyalkanoate (PHA) scaffolds with good mechanical properties and biocompatibility. Biomaterials 24:1041–1045
173. Ni J, Wang M (2002) In vitro evaluation of hydroxyapatite reinforced polyhydroxybutyrate composite. Mater Sci Eng C20:101–109
174. Chen DZ, Tang CY, Chan KC et al (2007) Dynamic mechanical properties and in vitro bioactivity of PHBHV/HA nanocomposite. Compos Sci Techn 67:1617–1626
175. Noohom W, Jack KS, Martin D, Trau M (2009) Understanding the roles of nanoparticle dispersion and polymer crystallinity in controlling the mechanical properties of HA/PHBHV nanocomposites. Biomed Mater 4:015003–015016
176. Kempen DH, Lu L, Kim Ch et al (2006) Controlled drug release from a novel injectable biodegradable microsphere/scaffold composite based on poly(propylene fumarate). J Biomed Mater Res 77A:103–111
177. Peter SJ, Lu L, Kim DJ et al (2000) Marrow stromal osteoblast function on a poly(propylene fumarate)/B-tricalcium phosphate biodegradable orthopedic composite. Biomaterials 21:1207–1213
178. Peter SJ, Miller ST, Zhu G et al (1998) In vivo degradation of a poly(propylene fumarate)/β-tricalcium phosphate injectable composite scaffold. J Biomed Mater Res 41:1–7
179. Peter SJ, Kim P, Yasko AW et al (1999) Crosslinking characteristics of an injectable poly(propylene fumarate)/B-tricalcium phosphate paste and the mechanical properties of the crosslinked composite for use as a biodegradable bone cement. J Biomed Mater Res 44:314–323
180. Jayabalan M, Thomas V, Rajesh PN (2001) Polypropylene fumarate/phloroglucinol triglycidyl methacrylate blend for use as partially biodegradable orthopaedic cement. Biomaterials 22:2749–2757
181. Heller J, Barr J, Ng SY et al (2002) Poly(ortho esters): synthesis, characterization, properties and uses. Adv Drug Deliv Rev 54:1015–1039
182. Heller J, Barr J, Ng SY et al (2000) Poly(ortho esters) – their development and some recent applications. Eur J Pharm Biopharm 50:121–128
183. Solheim E, Sudmann B, Bang G et al (2000) Biocompatibility and effect on osteogenesis of poly(ortho ester) compared to poly(DL-lactic acid). J Biomed Mater Res 49:257–263
184. Allcock HR (1976) Polyphosphazenes: New polymers with inorganic backbone atoms. Science 193:1214–1219
185. Allcock HR, Fuller TJ, Mack DP et al (1977) Synthesis of poly[(amino acid alkyl ester)phosphazenes]. Macromolecules 10:824–830
186. Nair LS, Lee DA, Bender JD et al (2006) Synthesis, characterization, and osteocompatibility evaluation of novel alanine-based polyphosphazenes. J Biomed Mater Res 76A:206–213
187. TenHuisen KS, Brown PW, Reed CS et al (1996) Low temperature synthesis of a self-assembling composite: Hydroxyapatite-poly[bis(sodium carboxylatophenoxy) phosphazene]. J Mater Sci Mater Med 7:763–782
188. Greish YE, Bender JD, Lakshmi S et al (2005) Low temperature formation of hydroxyapatite-poly(alkyl oxybenzoate)phosphazene composites for biomedical applications. Biomaterials 26:1–9

189. Greish YE, Bender JD, Lakshmi S et al (2005) Composite formation from hydroxyapatite with sodium and potassium salts of polyphosphazene. J Mater Sci Mater Med 16:613–620
190. Greish YE, Bender JD, Lakshmi S et al (2006) Formation of hydroxyapatite–polyphosphazene polymer composites at physiologic temperature. J Biomed Mater Res 77A:416–425
191. Krishnan M, Flanagan DR (2000) FTIR-ATR spectroscopy for monitoring polyanhydride/anhydride-amine reactions. J Control Release 69:273–281
192. Attawia MA, Uhrich KE, Botchwey E et al (1996) In vitro bone biocompatibility of poly(anhydride-co-imides) containing pyromellitylimidoalanine. J Orthopaedic Res 14: 445–454
193. Muggli DS, Burkoth AK, Anseth KS (1999) Crosslinked polyanhydrides for use in orthopedic applications: degradation behavior and mechanics. J Biomed Mater Res 46:271–278
194. Domb AJ, Mathiowitz E, Ron E et al (1991) Polyanhydrides. IV. Unsaturated and crosslinked polyanhydrides. J Polym Sci Part A Polym Chem 29:571–579
195. Anseth KS, Quick DJ (2001) Polymerizations of multifunctional anhydride monomers to form highly crosslinked degradable networks. Macromol Rapid Commun 22:564–572
196. Albertsson AC, Eklund M (1996) Short methylene segment crosslinks in degradable aliphatic polyanhydride: network formation, characterization, and degradation. J Polym Sci Polym Chem 34:1395–1405
197. Kumar N, Langer RS, Domb AJ (2002) Polyanhydrides: an overview. Adv Drug Deliv Rev 54:889–910
198. Bonzani IC, Adhikari R, Houshyar S et al (2007) Synthesis of two-component injectable polyurethanes for bone tissue engineering. Biomaterials 28:423–433
199. Bucky LP, Ehrlich HP, Sohoni S et al (1994) The capsule quality of saline-filled smooth silicone, textured silicone, and polyurethane implants in rabbits: a long-term study. Plast Reconstr Surg 93:1123–1131
200. Zhu Y, Gao C, He T et al (2004) Endothelium regeneration on luminal surface of polyurethane vascular scaffold modified with diamine and covalently grafted with gelatin. Biomaterials 25:423–30
201. Santerre JP, Woodhouse K, Laroche G et al (2005) Understanding the biodegradation of polyurethanes: from classical implants to tissue engineering materials. Biomaterials 26: 7457–7470
202. Guan J, Fujimoto KL, Sacks MS et al (2005) Preparation and characterization of highly porous, biodegradable polyurethane scaffolds for soft tissue applications. Biomaterials 26:3961–3971
203. Gorna K, Gogolewski S (2003) Preparation, degradation, and calcification of biodegradable polyurethane foams for bone graft substitutes. J Biomed Mater Res A 67:813–827
204. Grad S, Kupcsik L, Gorna K et al (2003) The use of biodegradable polyurethane scaffolds for cartilage tissue engineering potential and limitations. Biomaterials 24:5163–5171
205. Li C (2002) Poly(L-glutamic acid)–anticancer drug conjugates. Adv Drug Deliv Rev 54: 695–713
206. Polaczek J, Pielichowski J, Pielichowski K et al (2005) A new method of poly(aspartic acid) synthesis under microwave radiation. Polimery 50:812–820
207. Mann S (2001) Biomineralization: principles and concepts in bioinorganic materials chemistry. Oxford University Press, Oxford
208. Bigi A, Boanini E, Gazzano M et al (2004) Nanocrystalline hydroxyapatite–polyaspartate composites. Biomed Mater Eng 14:573–579
209. Sarig S (2004) Aspartic acid nucleates the apatite crystallites of bone: a hypothesis. Bone 35:108–113
210. Hinoi E, Takarada T, Yoneda Y (2004) Glutamate signaling system in bone. J Pharmacol Sci 94:215–220
211. Boanini E, Torricelli P, Gazzano M et al (2006) Nanocomposites of hydroxyapatite with aspartic acid and glutamic acid and their interaction with osteoblast-like cells. Biomaterials 27:4428–4433

212. Ha SW, Gisep A, Mayer J et al (1997) Topographical characterisation and microstructural interface analysis of vacuum-plasma-sprayed titanium and hydroxyapatite coatings on carbon fibre-reinforced poly(etheretherketone). J Mater Sci Mater Med 8:891–896
213. Blazewicz S, Chlopek J, Litak A et al (1997) Experimental study of mechanical properties of composite carbon screws. Biomaterials 18:437–439
214. Al-Shawi AK, Smith SP, Anderson GH (2002) The use of a carbon fibre plate for periprosthetic supracondylar femoral fractures. J Arthroplasty 17:320–324
215. Fujihara K, Huang ZM, Ramakrishna S et al (2004) Feasibility of knitted carbon/PEEK composites for orthopedic bone plates. Biomaterials 25:3877–3885
216. Wan YZ, Wang YL, He F et al (2007) Mechanical performance of hybrid bismaleimide composites reinforced with three-dimensional braided carbon and Kevlar fabrics. Composites Part A 38:495–504
217. Bacakova L, Grausova L, Vacik J et al (2007) Improved adhesion and growth of human osteoblast-like MG 63 cells on biomaterials modiefied with carbon nanoparticles. Diam Rel Mater 16:2133–2140
218. Chlopek J, Morawska-Chochol A, Paluszkiewicz C (2008) FTIR evaluation of PGLA – Carbon fibres composite behaviour under 'in vivo' conditions. J Molec Struct 875:101–107
219. Slosarczyk A, Klisch M, Blażewicz M et al (2000) Hot pressed hydroxyapatite-carbon fibre composites. J Eur Cer Soc 20:1397–1402
220. Dorner-Reisel A, Berroth K, Neubauer R et al (2004) Unreinforced and carbon fibre reinforced hydroxyapatite: resistance against microabrasion. J Eur Cer Soc 24:2131–2139
221. Piekarczyk-Rajzer I, Blazewicz M (2005) Carbon fibres modified with hydroxyapatite. Eng Biomater 47–53:63–65
222. Janowska G, Mikolajczyk T, Bogun M (2007) Effect of the type of nanoaddition on the thermal properties of polyacrylonitrile fibres. J Therm Anal Calorim 89:613–618
223. Zhao X, Zhang S (2006) Self-assembling nanopeptides become a new type of biomaterial, Adv Polym Sci 203:145–170
224. Itoh S, Kikuchi M, Koyama Y et al (2002) Development of an artificial vertebral body using a novel biomaterial, hydroxyapatite/collagen composite. Biomaterials 23:3919–3926
225. Rhee SH, Suetsugu Y, Tanaka J (2001) Biomimetic configurational arrays of hydroxyapatite nanocrystals on bio-organics. Biomaterials 22:2843–2847
226. Kikuchi M, Itoh S, Ichinose S et al (2001) Self-organization mechanism in a bone-like hydroxyapatite/collagen nanocomposite synthesized in vitro and its biological reaction in vivo. Biomaterials 22:1705–1711
227. Chang MC, Ikoma T, Kikuchi M et al (2002) The cross-linkage effect of hydroxyapatite/collagen nanocomposites on a self organization phenomenon. J Mater Sci Mater Med 13:993–997
228. Kikuchi M, Matsumoto HN, Yamada T et al (2004) Glutaraldehyde cross-linked hydroxyapatite/collagen self-organized nanocomposites. Biomaterials 25:63–69
229. Chang MC, Ikoma T, Kikuchi M et al (2001) Preparation of a porous hydroxyapatite/collagen nanocomposite using glutaraldehyde as a crosslinkage agent. J Mater Sci Lett 20:1129–201
230. Chang MC, Tanaka J (2002) XPS study of the microstructure development of hydroxyapatite-collagen nanocomposite crosslinked using glutaraldehyde. Biomaterials 23:3879–3885
231. Chang MC, Tanaka J (2002) FT-IR study for hydroxyapatite-collagen nanocomposite cross-linked by glutaraldehyde. Biomaterials 23:4811–4818
232. Zhai Y, Cui FZ (2006) Recombinant human-like collagen directed growth of hydroxyapatite nanocrystals. J Cryst Growth 291:202–206
233. Bhowmik R, Katti KS, Katti DR (2007) Mechanics of molecular collagen is influenced by hydroxyapatite in natural bone. J Mater Sci 42:8795–8803
234. Bernhardt A, Lode A, Boxberger S et al (2008) Mineralised collagen—an artificial, extracellular bone matrix—improves osteogenic differentiation of bone marrow stromal cells. J Mater Sci Mater Med 19:269–275
235. Wang Y, Cui FZ, Hu K et al (2008) Bone regeneration by using scaffold based on mineralized recombinant collagen. J Biomed Mater Res Part B Appl Biomater 86B:29–35

236. Teng SH, Lee EJ, Wang P et al (2008) Three-layered membranes of collagen/hydroxyapatite and chitosan for guided bone regeneration. J Biomed Mater Res Part B Appl Biomater 87B:132–138
237. Chang MC, Ko CC, Douglas WH (2003) Preparation of hydroxyapatite-gelatin nanocomposite. Biomaterials 24:2853–2862
238. Word AG, Courts A (1977) The science and technology of gelatin. Academic, London
239. Chiellini E, Cinelli P, Fernandes EG et al (2001) Composite materials based on gelatin and fillers from renewable resources. In: Chiellini E, et al. (ed) Biorelated polymers, Part II. biopolymer technology and applications. Kluwer, New York
240. Kim HW, Knowles JC, Kim HE (2004) Hydroxyapatite and gelatin composite foams processed via novel freeze-drying and crosslinking for use as temporary hard tissue scaffolds. J Biomed Mater Res A 72A:136–145
241. Kim HW, Knowles JC, Kim HE (2004) Porous scaffolds of gelatin–hydroxyapatite nanocomposites by biomimetic approach: characterization and antibiotic drug release. J Biomed Mater Res B 74B:686–698
242. Kim HW, Kim HE, Salih V (2005) Stimulation of osteoblast responses to biomimetic nanocomposites of gelatin–hydroxyapatite for tissue engineering scaffolds. Biomaterials 26:5221–5230
243. Shu C, Xianzhu Y, Zhangyin X et al (2007) Synthesis and sintering of nanocrystalline hydroxyapatite powders by gelatin-based precipitation method. Ceram Int 33:193–196
244. Teng S, Chen L, Guo Y et al (2007) Formation of nano-hydroxyapatite in gelatin droplets and the resulting porous composite microspheres. J Inorg Biochem 101:686–691
245. Rohanizadeh R, Swain MV, Mason RS (2008) Gelatin sponges (Gelfoam) as a scaffold for osteoblasts. J Mater Sci Mater Med 19:1173–1182
246. Hillig WB, Choi Y, Murtha S (2008) An open-pored gelatin/hydroxyapatite composite as a potential bone substitute. J Mater Sci Mater Med 19:11–17
247. Bera T, Vivek AN, Saraf SK et al (2008) Characterization of biomimetically synthesized Hap–Gel nanocomposites as bone substitute. Biomed Mater 3:025001
248. Catledge S, Tyagi P, Koopman M et al (2004) Electrospun gelatin/hydroxyapatite nanocomposite scaffolds for bone tissue engineering, mater. Res Soc Symp Proc 1094:1094-DD09–05
249. Vepari C, Kaplan DL (2007) Silk as a biomaterial. Prog Polym Sci 32:991–1007
250. Wong Po Foo C, Kaplan DL (2002) Genetic engineering of fibrous proteins: spider dragline silk and collagen. Adv Drug Deliv Rev 54:1131–1143
251. Altman GH, Diaz F, Jakuba C et al (2003) Silk-based biomaterials. Biomaterials 24:401–416
252. Nemoto R, Wang L, Ikoma T et al (2004) Preferential alignment of hydroxyapatite crystallites in nanocomposites with chemically disintegrated silk fibroin. J Nanopart Res 6:259–265
253. Kino R, Ikoma T, Yunoki S et al (2007) Preparation and characterization of multilayered hydroxyapatite/silk fibroin film. J Biosci Bioeng 103:514–520
254. Korematsu A, Furuzono T, Yasuda S et al (2005) Nano-scaled hydroxyapatite/polymer composite III. Coating of sintered hydroxyapatite particles on poly(4-methacryloyloxyethyl trimellitate anhydride)-grafted silk fibroin fibers. J Mater Sci Mater Med 16:67–71
255. Rinaudo M (2006) Chitin and chitosan: properties and applications. Prog Polym Sci 31: 603–632
256. Khor E, Lim LY (2003) Implantable applications of chitin and chitosan. Biomaterials 24:2339–2349
257. Wan ACA, Khor E, Hastings GW (1997) Hydroxyapatite modified chitin as potential hard tissue substitute material. J Biomed Mater Res 38:235–241
258. Kim IY, Seo SJ, Moon HS et al (2008) Chitosan and its derivatives for tissue engineering applications. Biotechn Adv 26:1–21
259. Di Martino A, Sittinger M, Risbud MV (2005) Chitosan: a versatile biopolymer for orthopaedic tissue-engineering. Biomaterials 26:5983–5990
260. Wan ACA, Khor E, Hastings GW (1998) Preparation of a chitin-apatite composite by in situ precipitation onto porous chitin scaffolds. J Biomed Mater Res 41:541–548
261. Ge Z, Baguenard S, Lim LY et al (2004) Hydroxyapatite–chitin materials as potential tissue engineered bone substitutes. Biomaterials 25:1049–1058

262. Zhao F, Yin Y, Lu WW et al (2002) Preparation and histological evaluation of biomimetic three-dimensional hydroxyapatite/chitosan-gelatin network composite scaffolds. Biomaterials 23:3227–3234
263. Zhao F, Grayson WL, Ma T et al (2006) Effects of hydroxyapatite in 3-D chitosan–gelatin polymer network on human mesenchymal stem cell construct development. Biomaterials 27:1859–1867
264. Li J, Chen Y, Yin Y et al (2007) Modulation of nano-hydroxyapatite size via formation on chitosan–gelatin network film in situ. Biomaterials 28:781–790
265. Zhang L, Li Y, Yang A et al (2005) Preparation and in vitro investigation of chitosan/nano-hydroxyapatite composite used as bone substitute materials. J Mater Sci Mater Med 16:213–219
266. Hu Q, Li B, Wang M et al (2004) Preparation and characterization of biodegradable chitosan/hydroxyapatite nanocomposite rods via in situ hybridization: a potential material as internal fixation of bone fracture. Biomaterials 25:779–85
267. Murugan R, Ramakrishna S (2004) Bioresorbable composite bone paste using polysaccharide based nano hydroxyapatite. Biomaterials 25:3829–3835
268. Liu H, Li H, Cheng W et al (2006) Novel injectable calcium phosphate/chitosan composites for bone substitute materials. Acta Biomaterialia 2:557–565
269. Kong L, Gao Y, Lu G et al (2006) A study on the bioactivity of chitosan/nano-hydroxyapatite composite scaffolds for bone tissue engineering. Eur Polym J 42:3171–3179
270. Chesnutt BM, Viano AM, Yuan Y et al (2009) Design and characterization of a novel chitosan/nanocrystalline calcium phosphate composite scaffold for bone regeneration. J Biomed Mater Res 88A:491–502
271. Kim S-H, Lim B-K, Sun F et al (2009) Preparation of high flexible composite film of hydroxyapatite and chitosan. Polym Bull 62:111–118
272. Araujo AB, Lemos AF, Ferreira JM (2009) Rheological, microstructural, and in vitro characterization of hybrid chitosan-polylactic acid/hydroxyapatite composites. J Biomed Mater Res 88A:916–922
273. Gomes ME, Ribeiro AS, Malafaya PB et al (2001) A new approach based on injection moulding to produce biodegradable starch-based polymeric scaffolds: morphology, mechanical and degradation behaviour. Biomaterials 22:883–889
274. Marques AP, Reis RL, Hunt JA (2002) The biocompatibility of novel starch-based polymers and composites: in vitro studies. Biomaterials 23:1471–1478
275. Mendes SC, Reis RL, Bovell YP et al (2001) Biocompatibility testing of novel starch-based materials with potential application in orthopaedic surgery: a preliminary study. Biomaterials 22:2057–2064
276. Klemm D, Schumann D, Udhardt U et al (2001) Bacterial synthetized cellulose – artificial blood vessels for microsurgery. Prog Polym Sci 26:1561–1603
277. Klemm D, Schumann D, Kramer F et al (2006) Nanocelluloses as innovative polymers in research and application. Adv Polym Sci 205:49–96
278. Rhee SH, Tanaka J (2000) Hydroxyapatite formation on cellulose cloth induced by citric acid. J Mater Sci Mater Med 11:449–452
279. Wan YZ, Huang Y, Yuan CD et al (2007) Biomimetic synthesis of hydroxyapatite/bacterial cellulose nanocomposites for biomedical applications. Mater Sci Eng C 27:855–864
280. Jiang LY, Li YB, Zhang L et al (2009) Preparation and characterization of a novel composite containing carboxymethyl cellulose used for bone repair. Mater Sci Eng C29:193–198
281. Kuo CK, Ma PX (2001) Ionically crosslinked alginate hydrogels as scafolds for tissue engineering: Part 1. Structure, gelation rate and mechanical properties. Biomaterials 22:511–521
282. Augst AD, Kong HJ, Mooney DJ (2006) Alginate hydrogels as biomaterials. Macromol Biosci 6:623–633
283. Zhang SM, Cui FZ, Liao SS et al (2003) Synthesis and biocompatibility of porous nanohydroxyapatite/collagen/alginate composite. J Mater Sci Mater Med 14:641–645
284. Stodolak E, Wator A, Woloska-Czapnik, D et al (2007) The effect of a fibrous phase on degradation of membrane composite materials for GBR/GRT application. Eng Biomater 10(61):16–20

285. Wang L, Li Y, Li C (2009) In situ processing and properties of nanostructured hydroxyapatite/alginate composite. J Nanopart Res 11:691–699
286. Prestwich GD, Marecak DM; Marecek, JF et al (1998) Controlled chemical modification of hyaluronic acid: synthesis, applications, and biodegradation of hydrazide derivatives. J Control Release 53:93–103
287. Mori M, Yamaguchi M, Sumitoma S et al (2004) Hyaluronan-based biomaterials for tissue engineering. Acta Histochem Cytochem 37:1–5
288. Solchaga LA, Dennis JE, Goldberg VM et al (1999) Hyaluronic acid-based polymers as cell carriers for tissue-engineered repair of bone and cartilage. J Orthop Res 17:205–213
289. Bakos D, Soldan M, Hernandez-Fuentes I (1999) Hydroxyapatite/collagen/hyaluronic acid composite. Biomaterials 20:191–195
290. Kim HD, Valentini RF (1996) Hyaluronan-based biodegradable scaffolds for sceletal tissue reconstruction. Proceedings of Fifth World Biomaterials Congress, Toronto, Canada, p 236
291. Shakesheff K (2006) Supports for tissue regeneration. Mater Today 9:63
292. Shirtliff VJ, Hench LL (2003) Bioactive materials for tissue engineering, regeneration and repair. J Mater Sci 38:4697–4707
293. Thomson RC, Shung AK, Yaszemski MY et al (2007) Principles of tissue engineering. Elsevier, San Diego
294. Shenoy AV (ed) (1999) Rheology of filled polymer systems. Kluwer, Dordrecht
295. Gao CY, Wang DY, Shen JC (2003) Fabrication of porous collagen/chitosan scaffolds with controlling microstructure for dermal equivalent. Polym Adv Technol 14:373–379
296. Kikuchi M, Ikoma T, Syoji D et al (2004) Porous body preparation of hydroxyapatite/collagen nanocomposites for bone tissue regeneration. Eng Mater 254:561–564
297. Yunoki S, Ikoma T, Tsuchiya A et al (2007) Fabrication and mechanical and tissue ingrowth properties of unidirectionally porous hydroxyapatite/collagen composite. J Biomed Mater Res Part B Appl Biomater 80B:166–173
298. Nalawade SP, Picchioni F, Janssen LP (2006) Supercritical carbon dioxide as a green solvent for processing polymer melts: processing aspects and applications. Prog Polym Sci 31:19–43
299. Wutticharoenmongkol P, Sanchavanakit N, Pavasant P et al (2006) Preparation and characterization of novel bone scaffolds based on electrospun polycaprolactone fibers filled with nanoparticles. Macromol Biosci 6:70–77
300. Welle A, Kroger M, Doring M et al (2007) Electrospun aliphatic polycarbonates as tailored tissue scaffold materials. Biomaterials 28:2211–2219
301. Greiner A, Wendorff JH (2008) Functional self-assembled nanofibers by electrospinning. Adv Polym Sci 219:107–171
302. Wahl DA, Sachlos E, Liu C et al (2007) Controlling the processing of collagen-hydroxyapatite scaffolds for bone tissue engineering. J Mater Sci Mater Med 18:201–209
303. Hollister JS, Maddox RD, Taboas JM et al (2002) Optimal design and fabrication of scaffolds to mimic tissue properties and satisfy biological constraints. Biomaterials 23:4095
304. Hollister SJ (2005) Porous scaffold design for tissue engineering. Nature Mater 4:518–524
305. Ma Z, Mao Z, Gao C (2007) Surface modification and property analysis of biomedical polymers used for tissue engineering. Coll Surf B Biointerfaces 60:137–157
306. Alves CM, Yang Y, Carnes DL et al (2007) Modulating bone cells response onto starch-based biomaterials by surface plasma treatment and protein adsorption. Biomaterials 28:307–315
307. Cole MA, Thissen H, Losic D et al (2007) A new approach to the immobilisation of poly(ethylene oxide) for the reduction of non-specific protein adsorption on conductive substrates. Surf Sci 601:1716–1725
308. Roach P, Eglin D, Rohde K et al (2007) Modern biomaterials: a review–bulk properties and implications of surface modifications. J Mater Sci Mater Med 18:1263–1277
309. Kim MS, Khang G, Lee HB (2008) Gradient polymer surfaces for biomedical applications. Prog Polym Sci 33:138–164
310. Kato K, Uchida E, Kang ET et al (2003) Polymer surface with graft chains. Prog Polym Sci 28:209–259
311. Wang DA (2007) Engineering blood-contact biomaterials by 'H-Bond Grafting' surface modification. Adv Polym Sci 209:179–227

312. Zhang S, Altman M (1999) Peptide self-assembly in functional polymer science and engineering. React Funct Polym 41:91–102
313. Sanghvi AB, Miller KPH, Belcher AM et al (2005) Biomaterials functionalization using a novel peptide that selectively binds to a conducting polymer. Nature Mater 4:496–502
314. Choi HW, Lee HJ, Kim KJ et al (2006) Surface modification of hydroxyapatite nanocrystals by grafting polymers containing phosphonic acid groups. J Coll Interface Sci 304:277–281
315. Lee JH, Lee HB, Andrade JD (1995) Blood compatibility of polyethylene oxide surfaces. Prog Polym Sci 20:1043–1079
316. Matyjaszewski K, Patten TE, Xia JH (1997) Controlled/living radical polymerization. Kinetics of the homogeneous atom transfer radical polymerization of styrene. J Am Chem Soc 119:674–680
317. Ando T, Kato M, Kamigaito M et al (1996) Living radical polymerization of methyl methacrylate with ruthenium complex: formation of polymers with controlled molecular weights and very narrow distributions. Macromolecules 29:1070–1072
318. Matyjaszewski K (2006) Controlled/living radical polymerization: from synthesis to materials. Oxford University Press, Oxford
319. Min K, Yu S, Lee H-I et al (2007) High yield synthesis of molecular brushes via ATRP in miniemulsion. Macromolecules 40:6557–6563
320. Goddard JM, Hotchkiss JH (2007) Polymer surface modification for the attachment of bioactive compounds. Prog Polym Sci 32:698–725
321. Jagielski J, Piatkowska A, Aubert P et al (2006) Ion implantation for surface modification of biomaterials. Surf Coat Techn 200:6355–6361
322. Ito Y, Hasuda H, Sakuragi M et al (2007) Surface modification of plastic, glass and titanium by photoimmobilization of polyethylene glycol for antibiofouling. Acta Biomater 3: 1024–1032
323. Liu H, Webster TJ (2007) Nanomedicine for implants: a review of studies and necessary experimental tools. Biomaterials 28:354–369
324. Hussain F, Hojjati M, Okamoto M et al (2006) Polymer-matrix nanocomposites, processing,manufacturing, and application: an overview. J Composite Mater 40:1511–1575
325. Cheremisinoff NP (1996) Polymer characterization: laboratory techniques and analysis. Noyes, New Jersey
326. Reading M (1993) Modulated differential scanning calorimetry – a new way forward in materials characterisation. Trends Polym Sci 1:248–253
327. Pielichowski K, Flejtuch K (2004) Phase behaviour of poly(ethylene oxide) studied by temperature modulated DSC. Influence of the molecular weight. J Macromol Sci, Part B – Phys B43:459–470
328. Okamoto M (2005) Biodegradable polymer/layered silicate nanocomposites: a review. In: Mallapragada S, Narasimhan B (eds) Handbook of biodegradable polymeric, materials and their applications. American Scientific, Valencia
329. Blazewicz M, Paluszkiewicz C (2001) Characterization of biomaterials used for bone tissue regeneration by FTIR spectroscopy. J Molec Struct 563–564:147–152
330. Price DM, Reading M, Hammiche A et al (1999) Micro-thermal analysis: scanning thermal microscopy and localised thermal analysis. Int J Pharm 192:85–96
331. Butt HJ, Cappella B, Kappl M (2005) Force measurements with the atomic force microscope: technique, interpretation and applications. Surf Sci Reports 59:1–152
332. Donaldson K, Stone V, Tran CL et al (2004) Nanotoxicology. Occup Environ Med 61: 727–728
333. Chan VSW (2006) Nanomedicine: an unresolved regulatory issue. Regul Toxic Pharm 46:218–224
334. Oberdörster G, Oberdörster E, Oberdörster J (2005) Nanotoxicology: an emerging discipline evolving from studies of ultrafine particles. Environ Health Perspect 113:823–839
335. Kagan VE, Bayir H, Shvedova AA (2005) Nanomedicine and nanotoxicology: two sides of the same coin. Nanomed Nanotechn Biol Med 1:313–316
336. Oberdorster G, Sharp Z, Atudorei V et al (2004) Translocation of inhaled ultrafine particles to the brain. Inhal Toxicol 16:437–445

337. Barnard AS (2006) Nanohazards: knowledge is our first defence. Nature Mater 5:245–248
338. Sanchez C, Arribart H, Giraud Guille MM (2005) Biomimetism and bioinspiration as tools for the design of innovative materials and systems. Nature Mater 4:277–288
339. Sahoo SK, Parveen S, Panda JJ (2007) The present and future of nanotechnology in human health care. Nanomed Nanotechn Biol Med 3:20–31

Editor: K. Dušek

Index

R

S

MIX
Papier aus verantwortungsvollen Quellen
Paper from responsible sources
FSC® C105338
FSC
www.fsc.org

If you have any concerns about our products,
you can contact us on
ProductSafety@springernature.com

In case Publisher is established outside the EU,
the EU authorized representative is:
Springer Nature Customer Service Center GmbH
Europaplatz 3, 69115 Heidelberg, Germany

Printed by Libri Plureos GmbH
in Hamburg, Germany